THE METROPOLITAN
TRANSPORTATION PROBLEM

THE METROPOLITAN
TRANSPORTATION PROBLEM

•

WILFRED OWEN

REVISED EDITION

THE BROOKINGS INSTITUTION

WASHINGTON, D.C.

Revised edition 1966
Charts by Cushing & Nevell
Second Printing June 1966

© 1956, 1966 by

THE BROOKINGS INSTITUTION
1775 Massachusetts Avenue, N.W., Washington D.C.

Library of Congress Catalogue Card Number 66-21151

THE BROOKINGS INSTITUTION is an independent organization devoted to nonpartisan research, education, and publication in economics, government, foreign policy, and the social sciences generally. Its principal purposes are to aid in the development of sound public policies and to promote public understanding of issues of national importance.

The Institution was founded December 8, 1927, to merge the activities of the Institute for Government Research, founded in 1916, the Institute of Economics, founded in 1922, and the Robert Brookings Graduate School of Economics and Government, founded in 1924.

The general administration of the Institution is the responsibility of a self-perpetuating Board of Trustees. The trustees are likewise charged with maintaining the independence of the staff and fostering the most favorable conditions for creative research and education. The immediate direction of the policies, program, and staff of the Institution is vested in the President, assisted by the division directors and an advisory council, chosen from the professional staff of the Institution.

In publishing a study, the Institution presents it as a competent treatment of a subject worthy of public consideration. The interpretations and conclusions in such publications are those of the author or authors and do not purport to represent the views of the other staff members, officers, or trustees of the Brookings Institution.

BOARD OF TRUSTEES

FOREWORD

The transportation problems of our cities and their rapidly expanding suburbs are the most urgent and neglected transportation problems confronting the country. In these heavily populated and industrialized areas, people are dependent on a complex system of transportation that is nearly everywhere in trouble. Obsolete facilities and growing demands have created seemingly insoluble difficulties, and present methods of dealing with these difficulties offer little prospect of relief.

In recent years metropolitan areas have borne most of the impact of population growth in the United States. This expanding population together with rising income is aggravating the strain on transportation facilities. As urbanization and economic growth continue, it is clear that major revisions in public policy will be necessary if traffic congestion is to be relieved. The problem cannot be solved merely by providing additional transportation capacity, however. A comprehensive approach must also consider intricate issues of public finance, urban government, urban planning, and the redevelopment of the patterns of urban living.

In this study the author has attempted to break new ground by looking at the problem of urban mobility as a whole and by exploring some of the interrelations between transport development and urban living. The objective is to determine the requirements of public policy demanded by modern transportation and modern cities in order to utilize to maximum advantage the benefits that each has to offer.

The Ford Foundation generously appropriated funds for the general support of the activities of the Brookings Institution, and this study was one of several made possible by that grant. For this support and for the freedom of research it affords, the Institution is deeply grateful.

The Institution wishes to acknowledge its indebtedness to

the many individuals who contributed to this study. Among these should be especially mentioned the Advisory Committee for the first edition, consisting of Charles L. Dearing, Carl Feiss, Walter P. Hedden, Frank W. Herring, Lewis H. Kimmel, and Robert B. Mitchell. Priscilla St. Denis assisted in the preparation of the study in her capacity as research assistant. The original manuscript was edited by Evelyn Breck and indexed by Ynid Robinson.

The task of updating the manuscript for the revised edition was accomplished with the skillful help of Inai Bradfield. The secretarial burdens were borne by Joan Canzanelli. The book is the product of the Economic Studies Division, which is under the direction of Joseph A. Pechman.

The views expressed in this book are those of the author and are not presented as the views of the staff members, officers, or trustees of the Brookings Institution.

ROBERT D. CALKINS
President

The Brookings Institution
January 1966

Important changes have taken place in the metropolitan areas of the United States since the first edition of this book was written ten years ago. Between 1955 and 1965 the number of urban residents in the United States increased 30 millions, and the number of motor vehicles increased 24 millions, to a total of 85 millions. Transit patronage declined by three billion rides, or 25 per cent. The results have raised the question whether it is possible to be urbanized and motorized and at the same time civilized.

To keep ahead of the mounting traffic jam, the nation has launched a major road-building effort. The most significant aspect has been the nationwide federal program of interstate highways. The undertaking includes 6,000 miles of urban expressways costing 13 billions of dollars. In addition, the federal government has embarked on a program to aid mass transit. At the same time, federal urban renewal efforts have multiplied public and private investment in slum clearance, redevelopment, and related urban construction. Also, in 1965 a new Department of Housing and Urban Development was created by the Congress, and in early 1966 the President recommended the establishment of a new Department of Transportation.

The nation has awakened to the task of building better cities, but as yet it is not fully prepared to tackle successfully the mounting problems that such an effort entails. Compared to ten years ago, little has been done to establish metropolitan or regional organizations necessary to create a desirable upgrading of the urban environment. In most communities the transport problem continues to be approached not as a system but as a series of separate highway, transit, and terminal problems. Progress has been made toward relating transport to community development, but accomplishments have been minor compared to the increasing need to make transport an integral part of improving urban living conditions.

On balance, the changes that have occurred have accelerated trends that were already a threat to urban living a decade ago. Important remedies have been initiated, but they have not been commensurate with the pace of urbanization and the deterioration of the urban scene. The present volume has updated the facts, but the conclusions of the earlier volume have not been substantially altered. The United States, considering its wealth, has made poor progress in the city toward improving either its standards of mobility or its standards of living.

WILFRED OWEN

January 1966

CONTENTS

CHAPTER I

The Metropolitan Transportation Dilemma

American cities have become increasingly difficult to live in and to work in largely because they are difficult to move around in. Inability to overcome congestion and to remove obstacles to mobility threaten to make the big city an economic liability rather than an asset.

The crisis in transportation is largely the result of the growing concentration of population and economic activity. In 1960 more than 125 million people were living in the cities and suburbs of the United States. Each year urban America is spreading at the rate of a million acres—an area as large as the state of Rhode Island. In the past decade and a half, the growth of urbanization has been equivalent to duplicating the populations of metropolitan New York, Detroit, Los Angeles, Chicago, and Philadelphia.

This concentration of people and resources in urban areas would have been impossible without the mobility and supply lines afforded by transportation. The capacity of the transport system and the low cost and dependability of transport services have enabled an increasing number of people to seek the economic, social, and cultural opportunities that urban living ideally provides. But paradoxically, metropolitan cities have now grown to the point where they threaten to strangle the transportation that made them possible.

The paradox is particularly striking because the past several decades have seen more revolutionary changes in transportation than all previous history. With the technical ability to solve its transportation problems well in hand, the modern city is confronted by a transportation problem more complex than ever before. Despite all the methods of movement, the problem in cities is how to move.

One reason for this dilemma is the fact that urban areas have been unable to adjust to the changing conditions brought about so rapidly by the technological revolution in transportation. The older urban centers, with physical characteristics that were fixed in less mobile times, have been staggered by the impacts of recent innovation. And the newer suburbs have compounded the transportation problem by duplicating the errors of downtown and by creating problems of public administration and finance that traditional governmental organization was not designed to meet.

The Problem and Its Impact

Every metropolitan area in the United States is confronted by a transportation problem that seems destined to become more aggravated in the years ahead. Growth of population and expansion of the urban area, combined with rising national product and higher incomes, are continually increasing the volume of passenger and freight movement. At the same time, shifts from rail to road and from public to private transportation have added tremendous burdens to highway and street facilities. They have created what appear to be insuperable terminal and parking problems. Continuing economic growth and the certainty of further transport innovation threaten to widen the gap between present systems of transportation and satisfactory standards of service.

Manifestations of the transportation problem in urban areas include the mass movement between work and home and the cost that it represents in money, time, and wasted energy. The transit industry is experiencing rising costs and financial difficulties, while the rider is the victim of antiquated equipment and poor service. Obsolescence and inadequate capacity have become characteristic of the highway network, and terminal problems mean high costs and delays for all forms of transportation. The speed of traffic in central business districts during so-called "rush hours" is frequently as low as six to ten miles an hour, and the problem is finding not only the room to move but a place to stop. The scattered location and obsolete design of freight terminals and the absence of satisfactory physical relationship among the several methods of transportation create a heavy volume of unnecessary traffic as well as delays and high costs that penalize business, the consumer, and the community.

For a nation with 85 million motor vehicles, relatively little has been accomplished toward adapting the city to the automotive age. A limited mileage of urban highways has been built to adequate standards, but for the most part traffic still moves on an antiquated gridiron of streets laid out long before the needs of the automobile were known. These streets were designed principally for convenient real estate platting and access to property rather than for mechanized transportation. Despite the congestion of city thoroughfares, the automobile and truck have been left to park haphazardly along the curb and to load and unload in the street where space is so badly needed for movement.

Highway standards are generally in inverse relation to the needs of traffic. The modern highway in open rural areas often degenerates in urbanized areas to an obsolete right-of-way crowded on both sides with commercial activities strung out in unsightly array to create what has been aptly called America's longest slums. In the city, the concentration of traffic on narrow streets with their numerous crossings means that the speed and service potentials of the motor vehicle cannot be realized. The accident toll is outrageous. Since the turn of the century, half a million people have been killed in motor vehicle accidents on city streets and millions of pedestrians have been injured.

The greatest transportation difficulties are experienced while commuting between home and work. The separation of housing facilities from employment centers together with the rapid expansion of the urban area have created a pendulum movement from home to work that accounts for a larger volume of passenger traffic than any other type of weekday travel. This movement is frequently accomplished with the most antiquated facilities and under the most frustrating conditions. The trip to work often cancels the gain from shorter hours on the job, and the daily battle with congestion is in sharp contrast to other improvements in modern working conditions.

Half a century of neglect has meant a long-term deterioration of transit service and failure to keep pace with technological change. Rising costs and declining patronage have led to a succession of fare increases and further reductions in service. In many cases, it has been impossible to set aside necessary allowances for depreciation of equipment, and the industry as a whole has been unable to attract sufficient capital to renew, modernize, or extend its services for the nearly eight billion riders per year who depend on public carriers.[1]

The cost of providing the physical facilities required to meet urban traffic requirements has reached astronomical levels. High costs of land and damage incident to construction and the tremendous capacity and complicated design of the facilities required in built-up urban areas have thus far combined to make a full-scale attack impossible. Fifty-two billion dollars spent for urban streets during the past four decades has been grossly inadequate to achieve a reasonable quality of transportation service. The cost and complexity of highway construction is indicated by the fact that some expressways cost from $10 million to $30 million per mile.

The contrast between these needs and the financial possibilities of meeting them is not indicative of easy solution. Many metropolitan cities in the United States are suffering from a chronic shortage of funds. Today nine tenths of the mounting expenses of city governments are for services that did not exist at the turn of the century—traffic engineering,

---

[1] The terms "transit" or "mass transportation" are synonymous and include surface street car, bus, or trolley bus in local urban service as well as rail rapid transit operating on exclusive rights-of-way, generally subway or elevated. The term "public transportation" includes transit or mass transportation plus rail commuter services and taxis.

airports, parking facilities, health clinics, and a long list of others. At the same time, every city is being overwhelmed with demands for better schools, housing, recreation facilities, and other public services along with improved transportation.

City governments burdened by the heavy outlays required to accommodate ever-growing volumes of city traffic frequently find that attempts to relieve congestion serve only to move the critical point somewhere else. Expressways or parking facilities established to meet the demand attract further use and magnify the need. Moreover, new facilities mean not only heavy capital outlays but the loss of large areas of land from the tax rolls, reducing receipts at the same time that added revenues are being sought.

Historical Nature of Urban Congestion

The urban transportation problem, although often thought of as relatively new and associated with the automobile and the United States, is both global and historic. All over the world the trend from agricultural economies to urban industrialization continues, and cities in every part of the globe are struggling with similar problems of achieving acceptable standards of urban mobility. Even where automobiles are few, the bus and truck and bicycle combine with less modern methods of movement to create a degree of chaos comparable to the least penetrable crosstown streets of New York.

The big city and its transportation problems were confounding the experts over a century ago, long before the complications of internal combustion. When the population of London increased from approximately one million in 1831 to over four million 60 years later, the poorer inhabitants of the city were forced to abandon the high-rent districts to commercial uses and the city was practically abandoned at night. Out of the resulting tide of traffic that ebbed and flowed from home to work came the commuter, or as he was more appropriately called in earlier times, the journeyman. One hundred years ago his oppression was experienced on foot, and a daily spectacle was "the streams of walkers two, three and four miles long, converging on the city."[2] But with the continued spread of the suburbs, the possibility of so solving the home-to-work problem became impractical and an ambitious program of railroad construction and bus operation was undertaken to cope with it. By the early 1890's a thousand London horse buses were carrying over 100 million passengers per year, and 400,000 daily commuters were carried into the city by rail.

Meanwhile American cities sought to relieve traffic congestion by constructing elevated and subway facilities. Surface transit vehicles were usurping so much street space in Boston more than 60 years ago that a subway was constructed to clear the way for other vehicles using the streets.[3] In 1905 congested traffic at rush hours was described as the number one problem of large cities in the United States, and pictures of urban traffic jams in the days of horse-drawn vehicles and electric cars attest to the fact that congestion was bad long before the motor vehicle made it worse. As early as 1902, the question was whether better results could be obtained "by starting on a bold plan on comparatively virgin soil than by attempting to adapt our old cities to our newer and higher needs."[4]

Although the urban transportation problem is both long-standing and world-wide, its characteristics are not everywhere alike. The problem varies widely among cities of different sizes, types, ages, and locations. Problems of a large metropolitan city are very different from those of a smaller town, and large cities themselves differ widely according to their history, topography, wealth, and function. But the long-standing nature of urban traffic congestion and its

[2] C. H. Holden, *The City of London, A Record of Destruction and Survival* (1951), p. 166. For a memorable picture of congestion at Ludgate Circus in 1870, see p. 67.

[3] Edward Dana, "Reflections on Urban Transit," An address before the Canadian Club, Montreal, April 21, 1947.

[4] Ebenezer Howard, *Garden Cities of Tomorrow* (1902), p. 134.

world-wide scope suggest, despite a variety of forms, that underlying factors may be universal and only partially related to modern methods of transport. Basic causes appear to be excessive crowding of population and economic activity into small areas of land and the disorderly arrangement of land uses that has maximized transport requirements. The great bulk and density of urban buildings and the concentration of employment in the downtown area have created a volume of passenger and freight movement that has become increasingly difficult to accommodate effectively regardless of transportation method. The congestion of people, horses, and street cars before the appearance of motorized transport, the rush-hour madness of the New York subways, and the lines of automobiles inching their way through the traffic circles of Washington are all manifestations of a continuing imbalance between transportation demand and available transport capacity.

Transportation and Urban Growth

Transportation has played a leading role in the congestion of cities. At an earlier time, heavy densities of population developed because the urban radius was limited to distances that could be covered on foot, or at best by horse. As lines of intercity communication were developed to serve the urban cores of the industrial age, they solved the problems of long-distance transportation that made it possible for great centers of production and employment to supply and support themselves.

But within the urban area, transport innovations were less successful in providing a better distribution of population and economic activity. Innovation itself could not assure the mobility that economic interdependence in an urban complex requires. Transportation facilities were designed primarily to carry people and goods into the center of the city where there were already too many people and an over-concentration of economic activity. Lack of transportation in the early stages of urban growth combined with the recent development of mechanical transport have created an urban environment in which "each great capital sits like a spider in the midst of its transportation web."[5]

The proportion of the population of the United States in urban places of all sizes has increased from 6 per cent in 1800 to 70 per cent in 1960. During the past decade, the spectacular growth of urban population has placed tremendous additional burdens on transportation in a short period of time.

From 1950 to 1960, when the population increased by 28 millions, 84 per cent of this growth took place in the nation's 212 metropolitan areas. The population of these areas increased by 26 per cent. The largest growth took place in suburban areas, which registered an increase of 49 per cent compared to the 11 per cent increase in central cities.[6] (See Chart 1.)

From 1960 to 1963, when there was a gain of 7.4 millions in the total population, 80 per cent of this took place in metropolitan areas. Population in these areas rose to 119 millions in 1963, and the growth rate of the suburbs was more than three times that of the central cities.[7]

The result of these trends has been to concentrate transportation problems in a relatively small number of metropolitan areas. From 1950 to 1960, the population of metropolitan areas increased 53 per cent in Los Angeles and

[5] Lewis Mumford, *The Culture of Cities* (1938), p. 323.

[6] The rate of population growth for the United States from 1950 to 1960 was 1.7 per cent per year for the country as a whole, less than 1 per cent outside metropolitan areas, and 2.3 per cent within these areas. Growth of the metropolitan areas themselves was 1 per cent in central cities and 4 per cent in the outlying areas. U. S. Bureau of the Census, "Growth of Metropolitan Areas in the United States: 1960 to 1963," *Population Characteristics,* Series P-20, No. 131, September 4, 1964, Table A.

[7] Estimates for 1963 from U. S. Bureau of the Census, "Growth of Metropolitan Areas in the United States: 1960 to 1963," *Population Characteristics,* Series P-20, No. 131, September 4, 1964, Table 1.

Houston, 45 per cent in Dallas, 35 per cent in Washington, D.C., 30 per cent in Seattle, 28 per cent in Minneapolis–St. Paul, and 25 per cent in Detroit.[8] In 1960 five metropolitan cities had more than three million people, 19 had between one million and three million, and 29 had between 500,000 and a million. The 24 most populous urban areas contained more than 60 million people. (See Appendix Table 2.)

POPULATION INCREASE, 1950 – 60

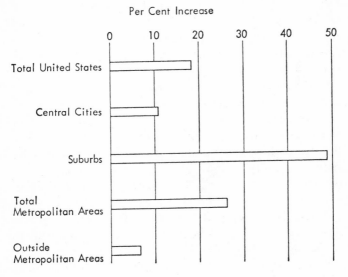

Based on Appendix Table 1.

Chart 1

The intensity of the transportation requirements in these urban places and the importance of urban transportation systems are indicated by relating the magnitude of city population to the postage-stamp areas that they encompass. Sixty per cent of the nation's people are located in 1 per cent of the nation's land. Thirty-three per cent of the popu-

[8] U. S. Bureau of the Census, *Census of Population: 1950* and *Census of Population: 1960*.

lation is concentrated in one-tenth of 1 per cent of the area of the country. The metropolitan area of New York City contains more people than the combined populations of Arizona, Delaware, Idaho, Maine, Montana, Nevada, New Hampshire, New Mexico, North and South Dakota, Rhode Island, Utah, Vermont, and Wyoming.

The high proportion of the population in the urban category and the small amount of land devoted to urban uses mean that the density of urban population is very high and consequently that the load on transportation facilities is very heavy. In 1960, there were 24,697 persons per square mile in New York City, and in Manhattan the figure was 77,000 per square mile. This compares with about 50 people per square mile for the country as a whole. In Chicago there were 12,959 persons per square mile, and in Philadelphia 15,743. There were 11 cities in the United States that had a population density of more than 11,000 per square mile. (See Appendix Table 3.)

Population figures do not measure the full magnitude of the transportation problem, for in addition to those who live in the city a large number of people come into the city during the daytime to work. This problem is indicated in the chart on page 11, which shows the resident and daytime populations of major cities in 1950. In 57 cities the daily flow of workers added another ten million people to the congestion at the center.[9] The daytime population of Boston, for example, was 34 per cent above the census figures of resident population, Pittsburgh had to accommodate 49 per cent more people during the day than at night, and in Newark, New Jersey, the population doubled during daylight hours.

The intensification of urban crowding that results from the daily influx of commuters is measured by the fact that whereas in 1950 only 4 per cent of the population in the five largest cities of the United States resided within a radius of one mile from the central business district, the estimated daytime population within a mile of the center

[9] Associated Universities, Inc., *Reduction of Urban Vulnerability, Report of Project East River*, Pt. V (July 1952), p. 14.

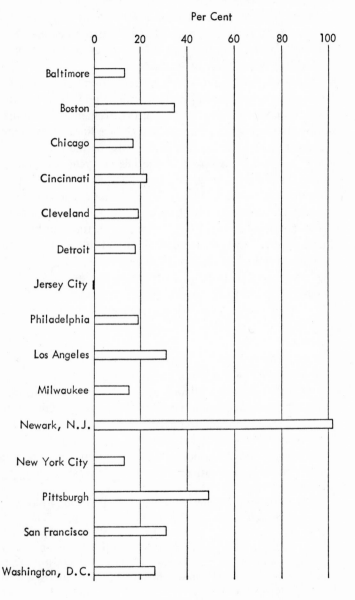

INCREASE OF DAYTIME
OVER RESIDENT POPULATION

Per Cent

Based on Appendix Table 4.

Chart 2

amounted to 30 per cent of the total population in each city. Less than 15 per cent of the residents of these largest cities lived within a two-mile radius from the centers, but estimated daytime population in that area amounted to half the resident population of each city.

To accommodate the heavy concentrations of people in urban centers, and to supply the factories and stores of the city with materials and goods requires a tremendous volume of movement under the most difficult space limitations. Supplying 100 million urban residents calls for the transportation of 2,000 million tons of materials per year. For every urban dweller an average of some 18 tons of materials is consumed annually.[10] How much the final consumer must pay to have the essential channels of mobility and supply kept open is not known, but the marvel is that the biggest cities are rarely inconvenienced by any visible break in the life lines on which they depend. The principal problems of which the average urban resident is aware are the inconveniences, discomforts, and exasperation of coping with the mounting obstacles to personal mobility.

City vs. Suburb

Relatively low-cost, reliable, and high-capacity transportation services have made possible these heavy concentrations of population and economic activity in the big city. The railroad lines to downtown, the subways, and the radial highways have supported congestion by creating the center and leading to it, and by making possible the fur-

[10] The President's Materials Policy Commission, *Resources for Freedom,* Summary of Vol. I (June 1952), p. 7. The task of feeding the city is illustrated by the collective appetite of New York City, whose daily diet consists of 2.5 million loaves of bread, 5 million quarts of milk, 15 million pounds of fresh fruits and vegetables, 4 million pounds of meat, and close to 30,000 gallons of wine. Every day 36,000 tons of refuse must be trucked away. Gilbert Millstein, "Statistics: Most of Them Superlatives," *New York Times Magazine* (February 1, 1953), p. 29.

nishing of supplies, the marketing of urban products, and the maintenance of minimum standards of mobility.

But more recently transportation has become an agent of dispersal as well, making possible the avoidance of concentration and promoting a diffused pattern of industrial and residential development. Symbolic of the new role of transport are the two-car family, the truck and bus, and the circumferential highway. The trend will continue with the impetus of vertical take-off aircraft, the heliport, and private travel by air. The problems of overcoming transportation difficulties are giving way to the possibilities of exploiting the advantages of transportation. The relative force of these two opposite aspects of transportation development will continue to play an important part in determining the character of urbanization in the future.

Currently, the most notable characteristic of urban change is the rapid growth of the fringes and the loss of population in central core areas. But there has been little evidence that declines in population or economic activity will be sufficient to diminish transport problems in the heart of the city in the near future. In New York City, for example, the population of Manhattan reached its peak in 1910 and declined thereafter until 1930. By 1950, there was an increase of 100,000 above the 1930 figure but between 1950 and 1960 population gradually declined by one quarter million. However, during this period Manhattan's loss was offset by an equal population increase in Queens.[11] The effect has not been to diminish the intensity of development close in but only to reduce the relative importance of New York City in the metropolitan area. Thus the fringe counties that contained only 8 per cent of the metropolitan population in 1910 accounted for 38 per cent in 1960, while the population of Manhattan declined in relative importance from 31 per cent of the metropolitan total in 1910 to 11 per cent in 1960.[12]

[11] The Port of New York Authority, *Metropolitan Transportation—1980*, New York, 1963, p. 347.

[12] The same, p. 348. Most of this shift had already taken place by 1930, however, when the figure was down to 16 per cent.

Similar trends in Chicago indicate that the high-density areas of the city are losing population very gradually. The resident population of Chicago within two miles of the cen-

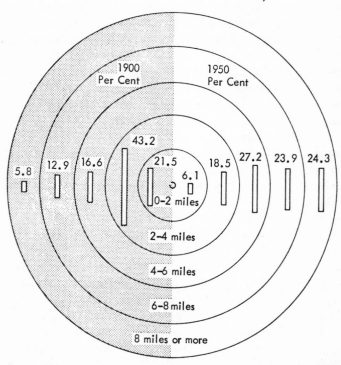

DISTRIBUTION OF CHICAGO POPULATION

(DISTANCE FROM CENTER OF CITY)

1900 Per Cent

1950 Per Cent

43.2

21.5

6.1

5.8

12.9

16.6

0-2 miles

18.5   27.2   23.9   24.3

2-4 miles

4-6 miles

6-8 miles

8 miles or more

Based on Appendix Table 5.

Chart 3

ter reached a peak in 1910, after which it declined sharply until 1940, then rose again.[13] The same pattern of change

[13] It may be assumed, however, that the growth recorded by the close-in areas during the decade of the 1940's was not a reversal of the long-run trend but was occasioned by the wartime

took place in the zone two to four miles from the center. Beyond the four-mile zone population has been increasing over a long period of time. The maximum rate of growth is taking place in the areas eight miles out and beyond. (See Chart 3.) [14]

In Philadelphia a century of population movement has not greatly reduced the high density of population within the city itself. In 1950, the number of people in the one- to two-mile zone was still 175 times greater than in the 18–25 mile zone. Within three miles of the city center there were almost a million people in 1950, the same as in 1900. The only difference was that with the spreading of the metropolitan area, total population within two miles of the center was a much smaller percentage of the metropolitan total compared to sixty years ago.[15] Urban redevelopment may be expected to result in further reductions in population density. The close-in areas of Philadelphia three to five miles from City Hall are expected to lose one fifth of their population by 1980.[16] Yet even with substantial losses

---

shortage of housing, which resulted in doubling up as well as the use of previously vacant dwellings.

[14] As a result of these trends, the proportion of the population of Chicago living within two miles of the center declined from 17 per cent in 1910 to 6 per cent in 1950; and the proportion living eight miles out or beyond rose from 9 per cent to 24 per cent. (See App. Tables 5 and 6.) Actually the number of people in the close-in zone was the same in 1930 and 1950, and the only substantial change from 1930 to 1950 was in the outer zone. *Growth and Redistribution of the Resident Population in the Chicago Standard Metropolitan Area,* A Report by the Chicago Community Inventory to the Office of the Housing and Redevelopment Coordinator and the Chicago Plan Commission (1954), p. 18.

[15] In 1950, 11 per cent of the total metropolitan population lived within two miles of the center compared to 24 per cent in 1900. Hans Blumenfeld, "The Tidal Wave of Metropolitan Expansion," *Journal of the American Institute of Planners,* Vol. 20 (Winter 1954), p. 13. See App. Table 7.

[16] Information from the Philadelphia Urban Traffic and Transportation Board.

there will still be a heavy concentration of people and economic activity.

For the majority of American cities, there will continue to be density reductions close in. These reductions plus the growth of the suburbs will further reduce the relative importance of the city itself, but substantial loosening up of the older congested areas will still leave heavy concentrations of urban population. Thus the urban area as a whole faces continuing problems of traffic congestion near the center along with the additional demands for suburban mobility and commutation service between suburb and center.

Employment trends parallel population trends. In New York City, for example, it is estimated that the decrease in the total number of jobs will be 64,000 in 1965 compared to the 1956 figure, while in the metropolitan area as a whole there will be an increase of 457,000 jobs. The proportion of total regional employment accounted for by New York City declined from 66 per cent in 1946 to 56 per cent in 1965, and thousands of people are now commuting from the city outward to get to work.[17] But there has been no visible decline in the magnitude of daily movement to the center.

Despite the spreading industrial growth in the suburbs, the city proper still retains a major share of the manufacturing activity in metropolitan areas. From the standpoint of the transportation problem of the city, growth trends have not diminished the factors contributing to traffic congestion. The kinds of industrial establishments moving out of the city are those with a relatively low number of employees per unit of area, whereas office work centering in the downtown area involves a high density of employment and the type of employment that creates the greatest peak-hour traffic.

As population and industry have grown in the suburbs,

[17] Regional Plan Association, Inc., *Population 1954–1975 in the New Jersey–New York–Connecticut Metropolitan Region,* Bulletin 85, p. 25. Also, Port of New York Authority, *Metropolitan Transportation—1980,* 1963, pp. 349–50.

retail business has expanded rapidly outside the city limits, and the percentage of metropolitan area sales being transacted in the city has declined. Again, however, the new stores have been necessary to accommodate the growth of population and income, and for the most part retail trade in the central city still flourishes.

Building permits for the year 1954 showed that half of the nation's construction took place in the suburbs of metropolitan areas and 31 per cent in central cities. The remainder was accounted for by nonmetropolitan areas.[18] These figures indicate that suburban growth has not destroyed the economic vitality in the central city. The tremendous suburban expansion has not appreciably altered the concentration of urban activity nor diminished the underlying causes of traffic congestion. What looked like dispersion a short while ago has been largely new growth, and this growth has of necessity taken place in the outer fringes where room for expansion is still available.

The Outlook for Urbanization

Despite the patterns of urban growth to date, and the presumption that centrifugal forces will gain rather than lose strength in the future, there is still no clear indication of the extent to which present trends in urban growth will continue. We do not know to what degree economic changes and developments in technology will alter the process of urbanization. Future growth may mean that existing densities in central cities will be substantially maintained or again increased; that conversely the downtown area and much of the central city will continue to give way before the onrush to the suburbs. Changes are also being introduced through planned cities of very different design that are making their appearance in many parts of the country.

Congestion and blight have multiplied the difficulties and frustrations of urban life, and in many places the growth

18 *Washington Post and Times Herald* (June 29, 1955).

of concentrated living seems to have passed the point of diminishing returns. Further vertical growth and urban sprawl both promise to compound the difficulties of providing transportation and other community facilities in the second half of the century. The threat of greater congestion has raised the question whether a nation born of farms is destined to die of cities.[19]

The belief that this could in fact be the case has been reinforced by a growing realization that until world peace can be guaranteed, the closely-packed city is particularly vulnerable to nuclear destruction. "A nation which keeps its wealth, its productive capacity, its population, and its administration huddled together in a few metropolitan areas, invites blackmail and courts disaster."[20] We are told on the one hand that "all the long-range dangers and disadvantages of our situation can be mitigated—even if none of them can be entirely eliminated—by judicious dispersal of our basic industries and productive population."[21] On the other hand, it is contended that the dangers of fall-out threaten to devastate so wide an area that to flee has become futile.

According to Frank Lloyd Wright, the deadline for decentralization has been so shortened by the threat of nuclear warfare that the urbanite must either be willing to get out of the city or be resigned to blowing up with it. Protection against enemy attack is no longer to be found in banding together in cities, and the possibilities of arriving at a more even distribution of the population over the unused areas of the country have been vastly increased by recent innovations. This is especially true of developments in transportation. In the light of these considerations alone, it may be that "further centralizations of any American city are only postponements of the city's end."[22]

Many of the economic advantages of urban living have

[19] Elmer T. Peterson, ed., *Cities Are Abnormal* (1946), p. v.
[20] "Must Millions March?", *Bulletin of the Atomic Scientists,* Editorial (June 1954), p. 194.
[21] The same, p. 195.
[22] Frank Lloyd Wright, "The Future of the City," *Saturday Review of Literature* (May 21, 1955), p. 12.

in any event already been neutralized. The city has become the victim of diseconomies that are reflected in high costs of living, including high costs of moving. Distribution costs and the difficulties of personal mobility often cancel other attractions of urban life. The theoretical benefits of urban location are frequently submerged in the rising discomforts and declining satisfactions of much that the city has to offer. The growing distances that must be covered from home to work, from one place of business to another, and from one friend to another tend to overcome the advantages of propinquity that the city is supposed to afford.

But more serious charges can be leveled at the urban area today. Blight and slums render large sections of the big city unfit for living. Although the wealth of the United States can be expressed in glowing statistics, about 20 per cent of the residential areas of its cities are slums. These slum areas contain one third of the urban population. To support them takes 45 per cent of the costs of municipal government, yet the same areas contribute only 6 per cent of total property tax revenues.[23] In New York City a quarter of a million dwellings lack toilet or bath, recreation facilities and open space are grossly inadequate, and even sunlight and air are at a premium.

Can we, then, provide a more satisfactory urban environment in new locations as transport technology enables us to move out and begin anew? The assumption that we can is often shaken by the newer suburbs. They have taken on many of the most objectionable aspects of the older blighted sections.[24] In attempting to flee from the undesirable conditions of the downtown area, frequently we have succeeded only in taking our mistakes with us.[25]

[23] National Association of Home Builders, *A New Face for America* (1953), p. 6. Data from Federal Works Agency, Public Buildings Administration.

[24] For a view of what is happening to "prosperous" suburbs see "Blight—Suburban Style," *Urban Land* (May 1955).

[25] "The far-flung metropolitan city of the motor age contains suburban slums and blighted commercial areas which are as appalling in their way as the old tight-packed city slums of the railway age." C. McKim Norton, "Metropolitan Transporta-

The idea of moving out to seek the health and enjoyment of air and sunlight has been a natural reaction to the noise and dirt of the city, but the endless spreading of cities has resulted in pushing the country farther and farther away. As a result, those who seek the restfulness or beauty of the countryside must constantly move outward to avoid the progressive waves of those who continue the escape from older blighted districts. This unplanned suburban development has resulted in the sprawl of large cities, the lengthening journey to work, and the growing difficulty of moving around.

Transportation has contributed in other ways to the diminishing desirability of urban living. The hazards and congestion of the highways, the noise and fumes of the motor vehicle, and the unsightliness of the gas station and used car lot have all added to the run-down character of the urban region. Transportation has created many of the conditions that people strive to escape, but it has also provided the means of escaping them and therefore the means of avoiding solutions. And it has transported slums to the suburbs.

The solution is not to abandon the city, therefore, but to assure that the inadequacies of the close-in urban area are corrected and that new suburban developments avoid the mistakes of the past. In some cities the congestion of the central city seems to have reached the saturation point, but we may also be arriving at an economic limit to the outward spread of the metropolis made possible by improved methods of transportation. As the increasing radius of travel resulting from modern transport permits us to move farther from the center, the cost of community facilities of all kinds increases. The higher densities of more compact urban areas of lesser size may offer a more economical alternative.

It can hardly be contended, however, that the degree of metropolitan concentration reached today is necessary for

tion," *An Approach to Urban Planning,* Gerald Breese and Dorothy E. Whitman, eds. (1953), p. 82.

the success of either business or the arts and sciences. Many of the activities that were necessarily located in the center of the city before the development of better means of transport and communications now could be decentralized and dispersed. The fact that some degree of concentration is an economic advantage does not lead to the conclusion that maximum concentration is the ultimate goal.

On the other hand, the big city can be defended on economic, social, and cultural grounds. The phenomenon of the metropolitan city derives from the fact that co-operative action makes possible greater productivity and higher standards of living, and it permits public services and amenities to be supplied more effectively and often at lower cost. A large labor force makes feasible the specialization required by large-scale industry and provides the skills needed for today's highly technical production processes. Even the availability of sufficient numbers of consumers to enjoy the fruits of modern production is predicated on the profitable markets afforded by concentrations of population. Advantages likewise stem from the variety of contacts and educational opportunities afforded by a larger population.

In any event, those who like the country have thus far been outnumbered by those who prefer a crowd, and the view that the city is bound to disappear has not as yet been borne out by events. More people are being attracted to cities and their urbanized surroundings than desire to remain rural, and at the center there continues to be a struggle between horizontal and vertical growth that has seen both sides claiming victory.

Alternative Transportation Solutions

What the future of the city will be or what the city of tomorrow ought to be like are questions closely related to the provision of transportation. Transport innovation will to a large degree dictate what is possible, and the extent to which transport policy is directed to achieving urban goals will help determine what is feasible. Many observers believe that a continuing downward trend in mass transportation is inevitable as car ownership expands and as highway and parking facilities are further developed to cope with traffic congestion. This would presumably hasten the decline of the center. Others take the view that in the relatively dense areas of the central city the attempt to accommodate the continuing trend toward private automobile transportation is a costly mistake that can end only in the ruination of downtown and the frustration of urban dwellers. The greater efficiency of mass transportation must be exploited, it is contended, by devoting more attention and money to the modernization and expansion of public conveyance, which in turn will preserve the downtown area.

One of the basic questions, then, concerns the relative emphasis to be placed on expressways and parking facilities to accommodate automobile use as compared to the modernization of mass transportation facilities aimed at restoring lost patronage and reducing the number of vehicles entering the city. If the latter course were followed, would it be possible to promote greater use of transit or would the urban resident either insist on using his private car or go elsewhere to work or shop or to do business? The correct decision is of basic importance to the future of the city and its people. The costs of transport modernization will be very high regardless where the emphasis is placed. But the cost of doing the wrong things or of simply doing nothing could be higher. For the ability to provide a circulatory system of acceptable standards will be an important factor in the economic survival of the urban economy.

The view that better mass transportation is the way out of the current situation is based on the contention that attempts to use the motor vehicle in an environment established before motoring needs were known are bound to be unsuccessful. Failure will be in the form of either downtown congestion or desertion. Mass transportation is capable of moving many times more people than automobiles can move, and under restricted space conditions should provide a more effective method of transportation. The problem of urban congestion has become so great that many communities are coming to the conclusion that there could never be sufficient highway and parking capacity to permit the movement of all people in private cars.

The opinion is frequently expressed that cities are suffering from "automobile blight"; that if the automobile were banned from downtown areas and satisfactory mass transportation provided instead, congestion would be relieved and greater freedom of movement would assure economic survival for the city. "The cities just cannot resign themselves to automobiles and let mass transportation slide to ruin and extinction. They must preserve mass transportation or stagnate."[26] Downtown is doomed to die, it is contended, unless cities stress movement of people rather than movement of vehicles. With this sentiment there appears to be widespread agreement. "Eventually cities will be places few people will want to live in, work in, or even visit unless they act to restrict private transportation."[27]

The mass transportation solution does not stimulate universal admiration, however. According to Mumford, while congestion originally provided the excuse for the subway, the subway has now become the further excuse for congestion. Small cities where people walked and rode bicycles were in a better position to take advantage of motor transport than cities that invested heavily in trolleys and rapid transit.[28]

If mass transportation is not the answer, what of the possibilities of modern highways to relieve the city of the congestion that inadequate transportation once made necessary? Critics insist that elaborate urban expressways are futile because of the tremendous reservoir of traffic waiting to absorb any new street capacity. According to this view,

[26] John Bauer, "The Crisis in Urban Transit," *Public Management* (August 1952), p. 176.

[27] W. H. Spears, quoting Joseph W. Lund in "Transit Is Dynamic," *Mass Transportation* (September 1954), p. 40.

[28] Mumford, *The Culture of Cities,* pp. 243, 441.

expressways and parking facilities not only will not solve the problem of congestion but will actually make it worse. The traffic engineer who tries to accommodate the private automobile "is doomed to inevitable failure . . . the better he does his job the greater will be his failure."[29]

But the position is also taken that the automobile, far from being a cause of urban congestion, has in fact made possible a necessary deconcentration of population through the decentralization of urban living and working. The endless streams of traffic that choke today's downtown streets make it natural to suppose that the private car has been responsible for the congestion of our cities, but it can be argued that the opposite is actually the case. "The only relief from congestion has been possible because of the motor vehicle."[30]

Still another view is that neither automobiles nor mass transportation nor any other mechanical contrivance can solve the problems of urban congestion. "As a solution of the traffic problem these devices are pure deception."[31] Putting the emphasis on supplying transportation facilities rather than controlling the demand, it is maintained, serves only to aggravate congestion. "As long as nothing is done fundamentally to rehabilitate the cities themselves, the quicker will people forsake them" and the greater the problems for those left behind to cope with.[32]

We have the assurance, therefore, that the problem of congestion in urban areas has been precipitated by the automobile; that the automobile, on the contrary, has been our escape from congestion; that the automobile and mass

[29] Walter Blucher, "Moving People—Planning Aspects of Urban Traffic Problems," *Virginia Law Review*, Vol. 36 (November 1950), p. 849.

[30] Arthur B. Gallion and Simon Eisner, *The Urban Pattern* (1950), p. 193.

[31] Mumford, *The Culture of Cities*, p. 296.

[32] Charles M. Nelson, "Expressways and the Planning of Tomorrow's Cities," in *Proceedings* of the Annual National Planning Conference, American Society of Planning Officials, Los Angeles, August 13–16, 1950 (1951), p. 121.

transportation are both guilty of promoting congestion; and finally that neither is the primary culprit, but rather a host of other factors that have resulted, thanks to modern technology, in the successful attempt to crowd too many people and too much economic activity into too little space. And of the city itself we are told that preservation of the vast investment in urban America will assure both economic salvation and nuclear annihilation.

Metropolitan areas thus face the difficult task of arriving at decisions that will determine to a major degree the physical and financial future of tomorrow's city. Should they emphasize expressways and parking facilities to accommodate automobile use, or modernize mass transportation facilities in the hope of restoring lost patronage and reducing the number of vehicles entering the city? Or will solutions depend instead on the extensive replanning and rebuilding of the American city? The next two chapters will explore the transportation problems of metropolitan areas and what is being done about them. This will set the stage for later exploration of a more comprehensive approach to urban mobility and its relation to urban finance, administration, and renewal.

CHAPTER II

Adapting to the Automotive Age

Motorized transportation has greatly increased the area of urbanization and the radius of travel for the urban resident. The time and distance limitations that once restricted the location of business and residential developments have been substantially removed. Urban growth that was once compelled to concentrate at the hub and along the rail spokes leading to the center is now free to spread in every direction along the roads that have become the giant skeleton of the new metropolis.

But the new mobility afforded by the automobile is often neutralized by obsolete highways, lack of parking, and by the high cost of providing the capacities and designs that safe and efficient motor vehicle use demands. In most large cities programs of highway development and traffic control accomplished to date have been unable to stem the flood of traffic. The volume of urban movement continues to outstrip the most ambitious engineering attempts to keep pace with demand. Financial policies and governmental machinery have been grossly inadequate to cope with the mounting intensity of the problem. All of these difficulties raise the question whether the attempt to meet the demands of the motor vehicle can succeed. Should the city adapt to the automobile or should transport technology instead be adapted to existing patterns of urbanization? To answer this question, the needs of automotive transportation will be analyzed in this chapter, and the problems of public carriers in the next.

Dependence on the Motor Vehicle

In the days of horse cars and pedestrians, a radius of between two and three miles was comfortable commuting distance for the urban dweller, and the size of most cities was limited to about 20 square miles. Transportation by street car, railroad, and rapid transit lines later expanded the radius of urban movement to five miles, and the urban area to 79 square miles. In the more recent automotive age, however, a commuting radius of 25 miles and more has become feasible, and the automobile has made possible an urban area covering upwards of 2,000 square miles.[1]

The most significant factor in the urban transportation picture today is that 80 per cent of all consumer units in the United States own a car.[2] As shown in the table on page 28, in many cities the proportion of car owners is much higher. In Columbus, Ohio, for example, 82 per cent of all families own one car and 24 per cent own more than one. In Fresno, California, 86 per cent of all families own one car and 33 per cent own two or more. In San Jose, California, the figures are 91 and 37.

This dispersal of both residential and industrial development made possible by the motor vehicle has resulted in a complex travel pattern. The largest number of people are still living and working in the city proper or working in the city and living in the suburbs. But many are now working in the suburbs and living in the city or both living and working in the suburbs. The sprawl of the urban area and the diffusion of trip origins and destinations has meant that in many locations the only feasible method of movement is by

[1] Harland Bartholomew, "Planning for Metropolitan Transportation," *Planning and Civic Comment,* Vol. 18 (September 1952), p. 1.

[2] Survey of Consumer Finances conducted by the Survey Research Center of the University of Michigan. Automobile Manufacturers Association, *Automobile Facts and Figures 1964,* p. 35.

automobile. Suburban shopping centers and industrial plants on the urban fringes have assumed that consumers and workers will come by car, and a high proportion of those who commute to the city find driving on congested highways preferable to the inconvenience and time loss of the trip by public carrier. Table 2 on page 29 shows the various ways that people get to work.

TABLE 1. Automobile Ownership in Selected Cities[a]

City	Percentage of Car-Owning Families	Percentage Owning More Than One Car
Chicago, Ill.	72	15
Philadelphia, Pa.–New Jersey	72	18
Boston, Mass.	73	16
Washington, D.C.–Md.–Va.	75	20
Pittsburgh, Pa.	76	16
St. Louis, Mo.–Ill.	77	18
Newark, N.J.	78	24
San Francisco–Oakland, Calif.	79	25
Cleveland, Ohio	80	23
Omaha, Neb.–Iowa	81	21
Wilmington, Del.–N.J.	82	22
Columbus, Ohio	82	24
Detroit, Mich.	83	25
Houston, Texas	83	30
Portland, Oreg.–Wash.	83	27
Honolulu, Hawaii	84	24
Dallas, Texas	85	31
San Diego, Calif.	86	30
Salt Lake City, Utah	86	31
Fresno, Calif.	86	33
Sacramento, Calif.	87	34
Phoenix, Ariz.	88	31
Flint, Mich.	89	24
Wichita, Kans.	89	30
San Jose, Calif.	91	37

Source: The data are derived from unpublished tables developed by Edmond L. Kanwit, Economic Research Divison, Bureau of Public Roads, "Selected Statistics by Standard Metropolitan Statistical Areas for Use in Transport Planning," August 7, 1964, Tables 3A, 3B, 3C.

[a] Data cover the entire metropolitan areas of the selected cities.

TABLE 2. Percentage of People Traveling to Work
by Different Modes of Transportation

Mode	Urban	Rural	Total U.S.
Automobile or car pool	64.2	63.3	64.0
Railroad (subway or elevated)	5.2	0.2	3.9
Bus or street car	10.9	1.0	8.2
Walked to work	10.0	9.7	9.9
Other means	2.2	3.4	2.5
Worked at home	2.9	18.9	7.2
Not reported	4.6	3.4	4.3
Total workers[a]	100.0	100.0	100.0

Source: Automobile Manufacturers Association, Automobile Facts and Figures, 1964, p. 40. Based on 1960 Census of Population, PC(SI) and PC(A3), Bureau of the Census, U. S. Department of Commerce.

[a] Includes armed forces.

The resulting dependence on privately owned vehicles for urban travel has resulted in the concentration of half of all motor travel in the United States in cities. An unknown but much higher proportion of automotive traffic takes place within the larger confines of metropolitan areas. Almost half of all commuters living within the city limits of metropolitan areas having populations of over one million travel by automobile. Within the city boundaries of San Francisco, 51 per cent of all commuters depend on automobile, while the comparable figure for San Diego and Minneapolis–St. Paul is 64. In New York City only about 20 per cent of commuters are automobile riders and in Boston the figure is 38, but this is a high proportion in relation to the size of the cities and the quality of available street accommodations. In smaller cities and suburban areas automobile use is much greater. For example, in cities with 200,000 to 250,000 people, 72 per cent travel by automobile, and the figure for cities of less than 100,000 is 76 per cent. (See Chart 4 and Appendix Table 8.)

For the nation as a whole, traffic volumes on city streets have more than doubled since 1940. The impact of this rapid growth has been felt particularly on the major urban

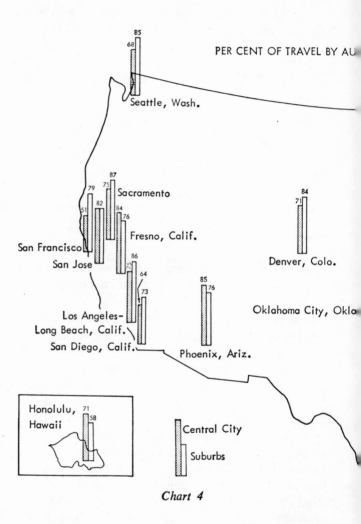

PER CENT OF TRAVEL BY AU

85
68
Seattle, Wash.

87
79 75 Sacramento
51 82 84
76
San Francisco
Fresno, Calif.
San Jose
86
64
73
Los Angeles-
Long Beach, Calif.
San Diego, Calif.
85
76
Phoenix, Ariz.

84
71
Denver, Colo.

Oklahoma City, Okla

Honolulu, 71
Hawaii 58

Central City

Suburbs

Chart 4

BILE IN SELECTED CITIES

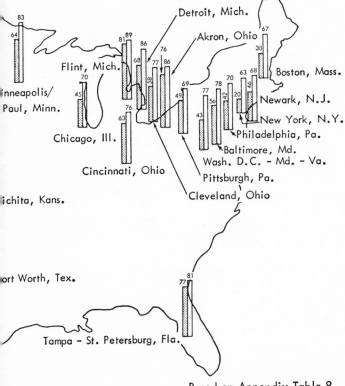

83
64

81 89
86
76
68 77 86
58

70
45

63

76

69
77 78 70
49 56 42 20

43

67
38

63 68
49

Detroit, Mich.

Akron, Ohio

Flint, Mich.

nneapolis/
Paul, Minn.

Chicago, Ill.

Cincinnati, Ohio

Pittsburgh, Pa.

Cleveland, Ohio

Boston, Mass.

Newark, N.J.

New York, N.Y.

Philadelphia, Pa.

Baltimore, Md.

Wash. D.C. – Md. – Va.

ichita, Kans.

ort Worth, Tex.

81
77

Tampa – St. Petersburg, Fla.

Based on Appendix Table 8.

TRANSPORTATION EXPENDITURE BY URBAN CONSUMERS

Per Cent of Total Expenditures

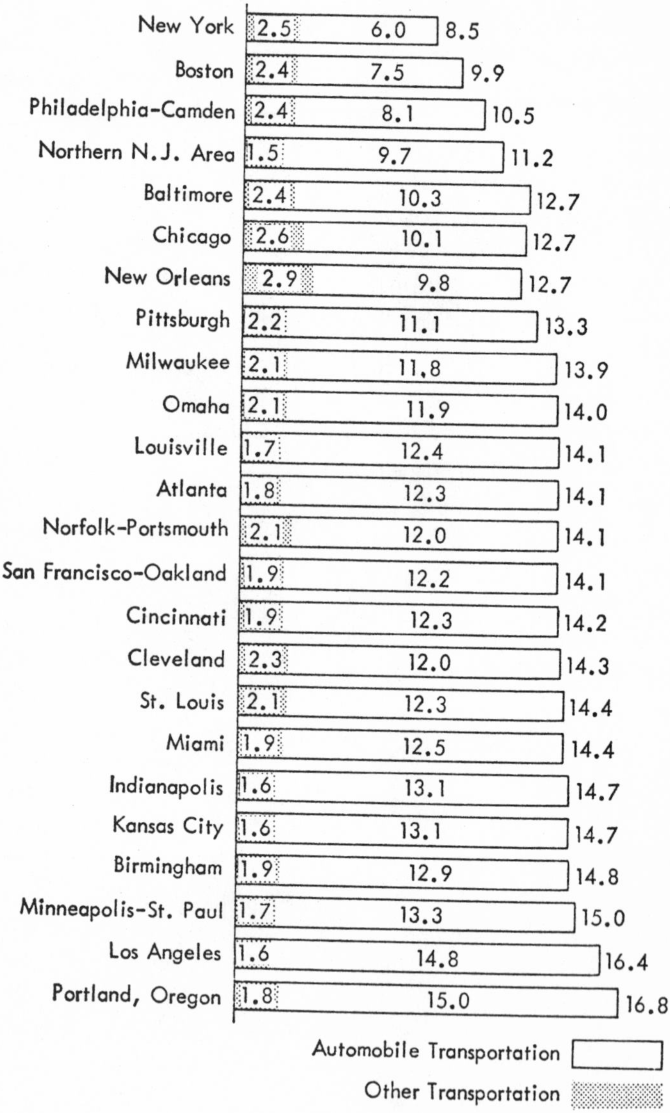

City	Other	Automobile	Total
New York	2.5	6.0	8.5
Boston	2.4	7.5	9.9
Philadelphia–Camden	2.4	8.1	10.5
Northern N.J. Area	1.5	9.7	11.2
Baltimore	2.4	10.3	12.7
Chicago	2.6	10.1	12.7
New Orleans	2.9	9.8	12.7
Pittsburgh	2.2	11.1	13.3
Milwaukee	2.1	11.8	13.9
Omaha	2.1	11.9	14.0
Louisville	1.7	12.4	14.1
Atlanta	1.8	12.3	14.1
Norfolk–Portsmouth	2.1	12.0	14.1
San Francisco–Oakland	1.9	12.2	14.1
Cincinnati	1.9	12.3	14.2
Cleveland	2.3	12.0	14.3
St. Louis	2.1	12.3	14.4
Miami	1.9	12.5	14.4
Indianapolis	1.6	13.1	14.7
Kansas City	1.6	13.1	14.7
Birmingham	1.9	12.9	14.8
Minneapolis–St. Paul	1.7	13.3	15.0
Los Angeles	1.6	14.8	16.4
Portland, Oregon	1.8	15.0	16.8

Automobile Transportation ☐
Other Transportation ▨

Based on Appendix Table 9.

Chart 5

arterials that on weekdays accommodate heavy commuter travel and on weekends provide the avenues of escape. The Hollywood Freeway in Los Angeles, opened in 1954, was designed to carry an ultimate future volume of 100,000 vehicles per day. It took only one year for traffic to reach the rate of 168,000 vehicles per day. Today the Harbor Freeway in Los Angeles serves 200,000 vehicles on an average day. In New York the George Washington Bridge carries a daily average of well over 100,000 cars, as does the Long Island Expressway. Chicago's Kennedy and Eisenhower Expressways carry between 150,000 and 200,000 vehicles per day, and the Lodge and Ford Expressways in Detroit account for even greater volumes. Several of San Francisco's highways carry well over 100,000 vehicles daily.[3]

Dependence on the automobile is indicated in the proportion of consumer expenditures being devoted to transportation. A study of Baltimore showed that the average family spent approximately $403 per year for automobile transportation, which was 10.3 per cent of its total consumption expenditures. Average family expenditures for local and intercity public carriers, however, totaled only $95, or 2.4 per cent of total annual consumer outlays. (See Chart 5, p. 32.)

These figures vary among different sizes and types of cities, but in all cases the automobile commands the major share of the transportation dollar. In New York it has been found that while only 8.5 cents out of each consumer dollar went for transportation, the automobile accounted for outlays of $294 per family per year compared to $121 for transit and intercity public carriers. In the more motorized city of Portland, Oregon, nearly 17 cents out of every consumer dollar was going for transportation, and the automobile was responsible for expenditure of $620 per family per year compared to $75 for local and intercity public carriers.

[3] Automotive Safety Foundation, "Urban Freeway Development in Twenty Major Cities," August 1964.

Prospective Growth of Automotive Transportation

Urban areas hoping for some slackening in the volume of automotive transportation and resulting traffic congestion will find little comfort in current growth trends. The basic factors affecting the outlook are the growth of urban living and automobile ownership that can now be anticipated. Current projections of population and long-term trends in car registrations seem to offer the ingredients of a traffic jam that could dwarf the difficulties experienced to date.

In the period 1950–60 the annual increase in population for the country as a whole was approximately three million per year. The probability that this average annual increase will at least be maintained over the next decade would mean a population of some 225 millions in 1975. The high projection by the Bureau of the Census indicates a population of 275 million in 1985.[4] (See Table 3.)

If the same proportion of the population were to live in urban areas in 1985 as is presently the case, urban dwellers that year would total 190 millions. But if the long-run trend toward urbanization were to continue, the number of people living in urban areas in 1985 might be much higher. If we assume, for example, that the proportion of urban to total population were to increase to 80 per cent of the total rather than remain at the present 69 per cent, urban dwellers would in that case number 221 millions in 1985.[5] This

[4] For purposes of planning physical facilities, the use of the high rather than the low or medium estimate offers the safest assumption.

[5] The trend in urban living, as indicated by data based on the old Census definition of urban areas, shows the proportion of population in urban areas increasing from 40 per cent in 1900 to 46 per cent in 1910; 51 per cent in 1920; 56 per cent in 1930; 57 per cent in 1940; and 59 per cent in 1950. According to the new Census definition, the proportion of urban to total population in 1950 was 64 per cent, and in 1960, 69 per cent. U. S.

would result in more urban residents that year than the entire population of the United States in 1965.

TABLE 3. Projection of Total and Urban Population to 1975 (In millions)

Midyear	Total U.S. Population[a]	Urban Population	
1950	151.7	97[b]	97[c]
1955	165.3	106	106
1960	180.7	125	125
1964	192.1	133	154
1970	211.4	146	169
1975	230.4	159	184
1980	252.1	174	202
1985	275.6	190	221

[a] U. S. Bureau of the Census, Population Estimate, Current Population Reports, Series P-25, No. 286, July 1964, as quoted in *Economic Report of the President,* 1965, p. 213. High estimates were used for the projected populations of 1970–85.

[b] The proportion of urban to total population was 64 per cent in 1950 and 1955. The percentage of urban population increased to 69 per cent in 1960 and for subsequent years assumption was made that the proportion of urban population remained at 69 per cent. The 1960 urban population figure is derived from U. S. Bureau of the Census, *Census of Population, 1960, Number of Inhabitants,* pp. 1–4. Urban population comprises all persons living in places of 2,500 inhabitants or more and the densely settled urban fringes around cities of 50,000 or more.

[c] Assuming a continuing increase in the proportion of urban population to total population, from 69 per cent to 80 per cent.

On the basis of these figures there might be anywhere from 57 to 88 million more people to be fed and supplied in urban areas two decades from now. The number of people requiring transportation to work will also increase as industrial employment rises to a probable 79 millions by 1975.[6] As a consequence, rush-hour traffic will be aug-

---

Bureau of the Census, *Census of Population: 1950,* Vol. 1, *Number of Inhabitants,* pp. 1–5, and *Census of Population: 1960.*

[6] U. S. Bureau of the Census, "A Projected Growth of the Labor Force in the United States Under Conditions of High Employment: 1950 to 1975," *Current Population Reports,* Series P-50, No. 42 (December 10, 1952).

mented by 20 million more urban workers than the number requiring transportation in 1960.

The likelihood is that automobile ownership will continue to increase much more rapidly than population in the next two decades. For growth of population will be accompanied by rising levels of economic activity, higher incomes, and a more equal distribution of personal income. These economic factors have made it possible for automobile ownership to increase by 25 million units in the short space of the first postwar decade.

As families move up the income ladder, increased expenditure for transportation always follows. This is what happened in the decade after the war, when per-capita income rose 40 per cent above the prewar level. In 1947 approximately 32 per cent of all consumer units had incomes of less than $3,000, but by 1963 only 19 per cent were in this low-income category. At the same time, the number of families in the relatively high income classes increased substantially.[7] The significance of this changing income distribution from the standpoint of transportation is indicated by the fact that only 53 per cent of families with less than $3,000 of annual income owned an automobile in 1963. But 95 per cent of families with incomes of $7,500 to $10,000 were car owners and 30 per cent of them owned two cars or more.[8]

A continuation of economic growth and prosperity, then, can be expected to mean further increases in automobile registrations. How much the increase is likely to be can be judged to some extent by past relations between car ownership and population. In 1962 there was one car for every

[7] U. S. Department of Commerce, *Survey of Current Business* (March 1955), p. 24. The proportion of families in the middle income groups receiving from $4,000 to $7,500 annual income rose from 22.8 per cent in 1944 to 39 per cent in 1953. See also, Council of Economic Advisers, *Economic Report of the President,* 1965, p. 162.

[8] Automobile Manufacturers Association, *Automobile Facts and Figures, 1964,* p. 38.

2.4 persons in the United States, compared to 4.8 persons per car in 1948. With an estimated population of 230 million people in 1975, and the same ratio of population to car ownership as today, the number of automobiles in operation that year would be approximately 81 millions compared to 70 millions in 1965.

It is apparent, however, that over the next decade better cars and better roads, along with economic growth, may be expected to alter the ratio in favor of a greater density of automobile ownership. Chart 6 indicates that the upward trend in automobile ownership has been closely comparable to trends in gross national product and much more rapid than population trends. Rising incomes and the growth of suburbs will also mean an increasing number of multiple-car families. The possibility of one car for every two people appears to be a conservative expectation and would mean a total of some 115 million cars by 1975. The probable addition of 20 million trucks would raise total 1975 registrations to the 135 million mark.

Obsolescence of Urban Facilities

These expectations of future motor vehicle ownership are a key to understanding the nature of the urban transportation problem that lies ahead.

With the mounting congestion of automotive traffic engulfing the cities, efforts to accommodate the motor vehicle have fallen far short of the need. Much of the street and highway system of the average city is obsolete in design, inadequate in capacity, and inefficient in operation. The urban areas where most vehicle owners live and where most of the driving is done have substantially reduced the potential advantages of the modern vehicle. The anticipated growth of population and car ownership outlined above can only mean that a very different approach will be necessary to meet the new needs of a motorized and urbanized economy.

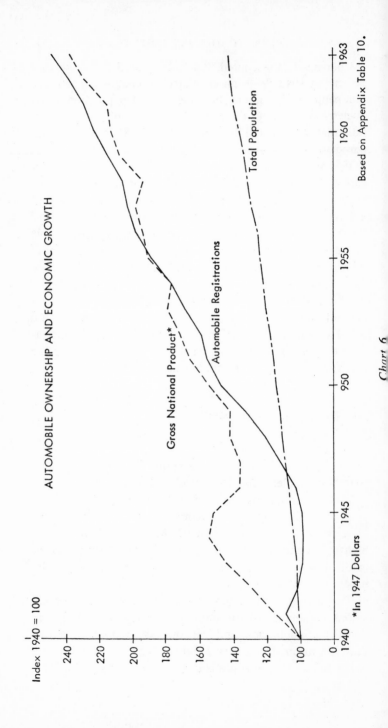

AUTOMOBILE OWNERSHIP AND ECONOMIC GROWTH

Index 1940 = 100

240

220

200

180

160

140

120

100

0

Gross National Product*

Automobile Registrations

Total Population

*In 1947 Dollars

1940   1945   1950   1955   1960   1963

Based on Appendix Table 10.

Chart 6

The automobile problem in urban areas derives principally from the fact that cities and their street layouts were designed for the most part before the requirements of the new vehicle were visualized. Even when the multiplication of motorized traffic made highway inadequacies obvious, the task of accomplishing more than superficial remedies seemed physically and financially out of the question.

The result has been that hard-pressed municipalities have to a large extent been forced to improvise solutions when only major programs of new construction could suffice to meet the need. But the existing road system is designed principally to bring traffic to the center, as was required before the greater flexibility of individual transport made such a traffic pattern obsolete. The result today is an absence of good circumferential distribution in the outer areas and a concentration of vehicles that cannot be accommodated in the downtown sections.

In suburban areas where new road building has progressed quite rapidly, the results have not been much different. Many miles of these highways have been encroached upon by commercial and residential uses that have reduced the capacity of the traveled way and increased the rate of traffic hazards. The disorderly strips of commercial development along the roadsides have at the same time created conditions of blight that have subtracted from the financial ability of the community to build modern facilities.

Most cities continue to settle for the unfortunate compromise of furnishing main highways to serve the dual purpose of moving traffic and providing access to land. These two functions cannot be supplied adequately by the same road. Numerous points of entrance and exit along the roadside and the traffic they generate seriously interfere with fast-moving vehicles and reduce both the capacity and safety of the highway. Hundreds of millions of dollars have been wasted in urban highway construction without controlled access despite ample evidence that such design is obsolete.

The death rate on conventional highways without control of access is generally two to four times as high as on roads

FATALITY RATES ON CONTROLLED ACCESS HIGHWAYS AND CONVENTIONAL ROADS

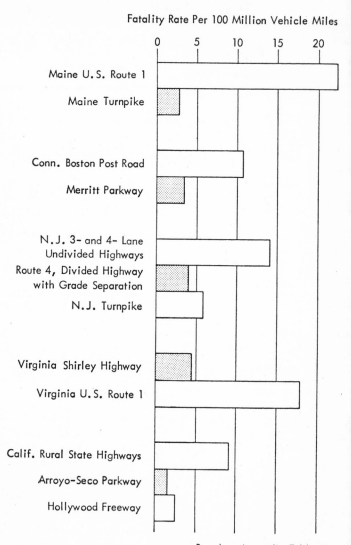

Fatality Rate Per 100 Million Vehicle Miles

Based on Appendix Table 11.

Chart 7

with access control.[9] On many routes the differential is greater. On U. S. Route 1 in Maine, for example, the fatality rate at the time the Maine Turnpike opened was 22.3 deaths per 100 million vehicle miles compared to 2.8 on the access-controlled Maine Turnpike. (See Chart 7.) Speeds of vehicle operation on controlled access facilities in urban areas is much greater than on ordinary streets. A summary of 12 case studies of highway speeds, shown in the accompanying table, indicates that with full control of access the speed of travel was 47 miles per hour both in cities and in rural areas. Without access control average travel speeds in urban places fell to 26 miles per hour, and in suburban areas to 39 miles per hour.

TABLE 4. Vehicle Speeds in Relation to Road Design[a]
(Miles per hour)

Road Design	Urban	Road Location Suburban	Rural
Full access control	47.3	49.2	47.4
Partial access control	—	42.3	49.5
No access control	26.4	38.9	44.9

[a] A. D. May, Jr., "Economics of Operation on Limited-Access Highways," *Vehicle Operation as Affected by Traffic Control and Highway Type*, Highway Research Board Bulletin 107 (1955), p. 55.

In addition to the effect of ribbon development in jeopardizing the highway investment, commercial properties strung along the highway right-of-way have caused the progressive deterioration of adjacent residential properties, the value of which has been reduced by bordering land uses that are frequently unsightly and generally incompatible with their surroundings. The expectation that roadside commercial establishments will ultimately eat into contiguous land as population growth creates the need for further expansion has discouraged property owners from undertak-

[9] *A Ten-Year National Highway Program*, A Report to the President, The President's Advisory Committee on a National Highway Program (January 1955), p. 9.

ing adequate maintenance and hastened the blight that uncontrolled strip development makes inevitable.

The worst roadside offenders in the American community today are the gas stations, used car lots, and other
establishments built to serve the motorist as part of the distribution and merchandising structure of the automotive industry. The paradox is that vehicle sales and service facilities contributing to congestion and hazards along the
traveled way are detracting from the effectiveness of
the highway system on which the industry depends for its
market. In addition, the automotive establishments that depend on the community for patronage are often the principal cause of the deterioration of the community. The
unsightliness of their places of business and the garish ar-
chitecture, signs, and lighting quickly reduce pleasant neighborhoods to blight.

The Cost of Modernization

It is these problems that caused the President of the
United States to warn the state governors in the summer of
1954 that the highway system had deteriorated to a point
where drastic action was necessary.[10] Some indication of
what would be required to meet the needs of motor traffic
had been provided by a joint survey made by the Commissioner of Public Roads in co-operation with the state highway departments.[11] This study, transmitted to the Congress
in the spring of 1955, reported that half of all urban street
mileage in the United States required improvement and
that one third of the needed improvements on the urban
portions of the federal-aid primary system involved the
construction of a complete new road on a new location.

[10] The President's message on "a 50 billion dollar highway
program in ten years" was delivered by Vice-President Nixon
to the Governors Conference at Bolton Landing, New York,
July 12, 1954.

[11] *Needs of the Highway Systems, 1955–84*, H. Doc. 120,
84 Cong. 1 sess. (1955).

Another 28,000 miles of new urban roads would be needed in the ten years from 1955 to 1964.[12]

TABLE 5. Urban Highway Requirements, 1955–84[a]
(In billions)

Urban System	Total First 10 Years	Total Next 20 Years	Grand Total
Interstate System	$10.8	$ 5.0	$15.8
Other Federal-Aid Primary	10.4	12.1	22.5
Total Federal-Aid	21.2	17.1	38.3
Other State Highways	2.0	2.3	4.3
Other Roads and Streets	18.4	30.7	49.1
Total Non-Federal Aid	20.4	33.0	53.4
Total, All Roads and Streets	41.6	50.1	91.7
Administrative Needs	2.0	2.4	4.4
Grand Total	43.6	52.5	96.1

[a] *Needs of the Highway Systems, 1955–84*, H. Doc. 120, 84 Cong. 1 sess. (1955), pp. 8, 14, 18, 19. Urban mileage, as applied to federal-aid systems, means mileage within and "adjacent to" municipalities or other urban places of 5,000 population or over. For the non-federal systems the states follow their own differing practices in classifying mileage as rural or urban.

It was estimated that between 1955 and 1985 total outlays required for highway construction, maintenance, and operation in the United States would amount to $297.1 billion. One third of this, or approximately $100 billion, would be in areas now designated as urban. (See Table 5.) In many states urban needs are a much higher proportion of the total than is indicated by national averages. In New Jersey 68.4 per cent of estimated highway construction needs were classified as urban, and in Rhode Island the figure was 58.8 per cent. Other states with more than 50 per cent of total highway needs in urban areas included California, Connecticut, Florida, and Illinois. Seven states reported urban needs at 40 to 49 per cent of the total.[13]

12 The same, pp. 11–12.
13 Louisiana, Maryland, Massachusetts, Minnesota, New York, Ohio, Oregon. Data from the U. S. Bureau of Public Roads.

Expressway costs in the period since the war far exceeded the most elaborate prewar requirements. This is shown in the table below. In Boston, three miles of the Central Artery cost $125 million. The first section of the elevated expressway, 1.7 miles long, required 2.7 miles of

TABLE 6. Cost of Urban Highways
I. Examples of Expressway Costs[a]

Expressway	Total Cost (In thousands)	Miles	Cost Per Mile (In thousands)
Hollywood Freeway	$ 55,000	10	$ 5,500
Arroyo Seco Freeway	11,000	8	1,375
John Lodge and Edsel Ford Expressway	207,000	24	8,625
Major Deegan Expressway	63,600	7.5	8,480
Cross-Bronx Expressway	112,000	5	22,450
Penn Lincoln Parkways	150,000	20	7,500
Boston Central Artery	125,000	3	41,667
Congress Street Expressway	50,000	8	6,250
Schuylkill Expressway	80,000	17	4,705

II. Average Cost Per Mile[b]
(In thousands)

Road System	Average Cost Right-of-Way	Total	Range of Cost Low 10 Per Cent	High 10 Per Cent
Interstate, urban:				
4-lane	$ 359	$1,414	$ 845	$ 2,250
6-lane	1,208	3,896	2,300	6,200
Over 6-lane	2,453	7,819	4,600	12,500
Federal-aid primary urban:				
4-lane	211	789	410	1,450
6-lane	810	2,312	1,425	3,760
Over 6-lane	1,000	2,819	1,600	4,960
Other city streets	35	140	57	300

[a] *Engineering News-Record* (February 17, 1955), pp. 124–30; Paul O. Harding, "Southern Freeways," *California Highways and Public Works* (January-February 1954), p. 13.
[b] U. S. Bureau of Public Roads, *Highway Needs 1955 to 1964. All Roads and Streets* (October 1954). Sec. D., pp. 4–5.

ramps. Total cost was $57 million. Ten miles of the Holly-
wood Freeway in Los Angeles required outlays of $55 mil-
lion, and eight miles of the Congress Street Expressway in
Chicago cost $50 million. In Detroit, the first 24 miles of
the expressway system cost over $200 million. A third tube
of the Lincoln Tunnel in New York cost $3 million more
than the original pair of tubes with their approaches. The
five-mile Cross Bronx Expressway involved a cost of over
$22 million per mile, and an eight-mile connection between
the New Jersey Turnpike and the Holland Tunnel, including a high-level bridge across Newark Bay, cost $114 mil-
lion.

In the city of Elizabeth, the acquisition and clearing of
right-of-way for the New Jersey Turnpike involved the removal or demolition of some 240 buildings, and relocation
of public utilities along the route cost over $8 million.[14] In
acquiring right-of-way for the Hollywood Freeway in Los
Angeles, it was necessary to demolish 90 buildings and
to move 1,728 others. The cost of right-of-way alone for
the Los Angeles Harbor Freeway was $10 million for a sin-
gle mile in the downtown area. Relocation of public utilities
along the 22.8 mile route cost $2.3 million.[15]

Beginnings of the Expressway Systems

Despite these costs and other obstacles, the 1950's saw
the beginning of many successful urban expressway programs. Expressway construction in the Boston metropolitan
area, for example, undertook to provide an extensive system of inner and outer circumferential highways and connecting radial expressways. The program was based on an
87-mile master highway plan for the metropolis.[16]

[14] New Jersey Turnpike Authority, *Annual Report 1950*, pp.
52, 57.
[15] W. L. Fahey, "Rapid Progress on Harbor Freeway," *California Highways and Public Works* (May–June 1954), pp. 3,
15.
[16] Joint Board for the Master Highway Plan, *Master Highway
Plan for the Boston Metropolitan Area* (February 1, 1948).

The presently planned freeway system for the urbanized area consists of three loops and eight radial expressways. The three loops are the inner belt (of which the Central Artery is a part), Route 128 surrounding the city of Boston, and the outer belt, Interstate Route 495.

Route 128 is a 65-mile circumferential highway built in rural surroundings ten miles from the central business district. This road provides easy access to a large number of communities in the metropolitan area as it circles the city. The most impressive result of the Route 128 project has been the stimulation of new industrial development. Although the entire road was constructed outside the urbanized area through agricultural land, the work had hardly been completed before great industrial plants began to mushroom on open grazing lands. This area, despite its proximity to the city, had defied developers for 300 years. The new establishments consist of light industry that can be served by truck and to which workers can drive from nearby cities and towns such as Concord, Lexington, Waltham, Wellesley, and Newton.[17] Traffic demand has required that a major portion of Route 128, originally built as a four-lane facility, be reconstructed to eight-lane width after only ten years of service.

The unique characteristic of this development is the uniform standard of design, set-back, and landscaping accomplished through one developer. More than 40 plants costing $100 million were under way before the road was finished, raising land values along the route from between $50 and $100 per acre to $2,000 to $5,000 an acre.[18]

The outer belt being built at a distance of 25 miles from the central business district, extends through a number of highly developed cities and towns surrounding the Boston area and reaches nearly to New Hampshire on the north

[17] Industrial and research firms locating along the route include such companies as American Can, CBS Hytron, Sylvania Electric, Polaroid, Arthur D. Little, Tracerlab, Vectron, and Raytheon.

[18] *Business Week* (May 14, 1955), pp. 186–90.

and Rhode Island on the south. The entire length of the outer belt lies outside the Boston urbanized areas.

The expressway system serving Boston was initiated by state legislative authorization for a bond issue of $100 million to be retired by receipts from the current tax on gasoline. Of this amount 45 per cent was allocated to the Boston metropolitan area. Subsequent bond issues during the period 1950 to 1954 brought the total authorized borrowing to $550 million. Gasoline taxes were raised two cents a gallon to service the debt.[19] No federal aid was used on the project.

The most ambitious program of urban expressway development in the United States has been undertaken in the Los Angeles area. The tremendous pressure of traffic gave rise to a freeway plan comprising 613 miles[20] of grade-separated highways in the metropolitan area, 290 of which are within the city of Los Angeles. It is estimated that the entire system will cost over $2 billion.[21]

The Los Angeles freeway system was developed by city, county, and state governments, and financed entirely out of current revenues. Funds were made available from a share of the regular state gasoline tax plus the proceeds of a gasoline tax imposed by the state legislature for the express purpose of building freeways. California law made the state responsible for the construction and maintenance of all state highways through cities on the same basis as all other state highways. Only a small mileage of the routes of the over-all plan were not in the state highway system. The cities and Los Angeles County, however, have contributed to the construction of some of the street extensions and improvements necessary to provide access to the freeways.[22]

[19] John McCloskey, "The Expressways Program of Massachusetts," Paper presented at ARBA Convention, Boston, February 1953.

[20] By 1964, the system comprised 769 miles.

[21] Paul O. Harding, "Southern Freeways," *California Highways and Public Works* (January–February 1954), p. 2.

[22] G. T. McCoy, "California Is Attacking Its Critical Deficiencies," *Better Roads* (June 1954), p. 24.

The dilemma of Los Angeles is that traffic continues to outstrip the rapid pace set by the road builders. During rush hours the tremendous jam on the major freeway routes has reduced average speed and intensified accident hazards. The wide dispersion of origins and destinations and the tremendous area of the metropolis make the provision and use of public transportation service extremely difficult. Yet the nerve-racking negotiation of the freeways in rush hours is an experience that many motorists find increasingly unpleasant, and the question is whether Los Angeles has built too many accommodations for the automobile or simply not enough.

The Detroit metropolitan area, which is heavily dependent on the automobile outside the central business district, has undertaken an extensive road-building program in an effort to provide a badly needed system of expressways. Local officials took the initiative and were given comparatively generous state support. The plan called for approximately 110 miles[23] of expressways, mostly depressed, with cross streets carried over the main artery. The sections completed have already made an imposing impact. They have stepped up the speed and comfort of urban travel and have opened up large areas of the city that were previously congested with commercial and housing development.

The Detroit system was initiated by the issuance of bonds secured by an agreement among state, county, and city, which makes specific sums of motor vehicle tax revenue available each year to match federal aid. Originally, the attempt had been made to build the expressways on a pay-as-you-go basis by an agreement pledging $1.5 million annually from the city of Detroit, an equal amount from the Wayne County Road Commission, and $3 million from the State Highway Department. This money, matched by urban federal aid, provided about $10 million per year. On this basis it would have taken 17 years to complete the first two expressways.

It was obvious, therefore, that a bonding program would be necessary, and state legislation was passed permitting the

[23] As of 1964, the proposed system comprised 329 miles.

state, county, and city to pledge a portion of their anticipated share of highway revenues for the retirement of limited access highway bonds. This solution was possible because substantial allocations of user revenues were being made to Michigan cities under existing law, and because the Michigan State Highway Department was required to spend 40 per cent of its construction funds in urban areas. These two favorable factors meant that substantial funds could be counted on each year to service a bond issue.

With debt service requirements thus provided for, a state constitutional amendment proposed by the city of Detroit was adopted to permit the issuance of "dedicated revenue bonds" for expressways. The financial plan was completed by amendment to the Federal-Aid Highway Act, providing that federal appropriations could be used to retire highway bonds.[24] The new tripartite bonding agreement substantially reduced current revenue requirements and made it possible for the first 25 miles of expressway to be completed in five years instead of 17.[25]

In the superhighway program for Chicago the most widely publicized section of the Chicago plan was the eight-lane Congress Street Expressway.[26] This highway is designed to accommodate rapid transit in the central mall. The road was financed by city, county, and state-federal funds. Various sections of the facility were built by each of the several state and local governments involved, and each assumed one third of the total cost of right-of-way and construction. The city of Chicago and Cook County both used their share of state motor fuel taxes to pay for part of the program, with the balance taken care of by municipal and county bonds.[27]

[24] Federal-Aid Highway Act of 1950, 64 Stat. 785, Sec. 5.

[25] Information from Glenn C. Richards, Commissioner of Public Works, Detroit.

[26] The name has since been changed to Eisenhower Expressway.

[27] Hugo C. Duzan, William R. MacCallum, Thomas R. Todd, "Recent Trends in Highway Bond Financing," *Public Roads* (October 1952), p. 83.

In addition to the intergovernmental arrangements for expressway projects concluded in some cities, ambitious and costly motor vehicle facilities have been undertaken in other areas through toll financing. Reliance on toll facilities for bridges and tunnels has been greatest in the New York area. Elsewhere in the United States, Boston, Philadelphia, and San Francisco have made use of toll financing. The Golden Gate and Bay bridges in San Francisco have demonstrated the financing possibilities of the toll method in that area, and the Sumner Tunnel and Mystic River Bridge have been successful projects in Boston. The Delaware River Bridge Authority has made possible the crossing between the Pennsylvania and New Jersey sections of the Philadelphia metropolitan area.

A number of urban toll highways have also been constructed. The eight-mile Calumet Expressway in Chicago provides access from the western end of the Indiana Toll Road to the center of Chicago. In addition, a number of existing rural toll roads have been extended into urban areas. The eight-mile Newark Bay–Hudson County extension of the New Jersey Turnpike connecting with the Holland Tunnel was constructed by the Turnpike Authority at a cost of nearly $15 million per mile. Other urban toll facilities include the New England section of the New York Thruway of 15 miles, which extends from within the Bronx through densely built-up sections of Westchester County to the Connecticut line. Another section of the Thruway penetrates downtown Buffalo. The Massachusetts Turnpike extension into Boston carries the expressway from the western suburbs into the heart of the city. In Virginia, the 34-mile Richmond-Petersburg Turnpike passes through both cities and provides 17 interchanges to serve predominantly local traffic.

Important progress is also being made toward developing the legal and financial tools necessary to provide adequate rights-of-way for urban highways. One innovation is the revolving fund for land purchase, which makes it possible for communities to acquire highway rights-of-way for future needs as the opportunity arises. Money from the revolving fund is used until an appropriation is made for actual construction. At that time the money already spent for the land is included in the appropriation and returned to the revolving fund for further land purchases. In California a land purchase fund of this type, to be used exclusively for highways, made it possible in a three-year period to acquire with $19 million rights-of-way that would have cost $114 million if the transactions had been postponed until construction was ready to start.[28] The use of this device on a metropolitan basis for all land requirements could eliminate the present situation in which open land needed for eventual public use either disappears or skyrockets in value.

The reservation of rights-of-way for future highway development is another successful device for overcoming land acquisition difficulties. This tool has been used in a number of cities and states, and involves no expenditures of public funds. Under the police power of the state, it is possible to prevent any permanent structures from being erected along locations officially designated as the site of a future highway. Development of the land in question is frozen with the official recording of the route. Purchase of the land can then be postponed until it is actually needed, when it is paid for at the prevailing price. By reserving the right to make the purchase, however, it is possible to prevent development that would either prohibit the highway project altogether or subject the public to the excessive cost of paying for the buildings and damages involved in the land purchase.

Role of the Federal Government

The role of the federal government in the urban highway program has been of increasing importance since the

[28] David R. Levin, "Highway Right-of-Way Acquisition and Control Problems," Address before the American Public Works Association Meeting, Milwaukee, Wis., October 5, 1955, p. 10 (mimeo.). Many other states have established similar revolving funds.

passage of federal-aid legislation in 1944, and particularly since the act of 1956. But hindsight makes it clear that federal policy makers should have acted sooner. For many years the federal government denied any concern for urban highways. Federal highway aid, originally administered through the Department of Agriculture, was strictly a rural program. The Federal-Aid Road Act of 1916 provided that no federal funds could be spent in cities of more than 2,500 population "except that portion of any such street or road along which the houses average more than two hundred feet apart."[29] In the 1930's federal help was given to cities for highway work, but the primary objective was to provide employment. It was not until 1944 that federal-aid legislation specifically authorized federal funds for urban streets.

The Federal-Aid Highway Act of 1944 called for the designation of a federal-aid highway system in urban areas and allocated $125 million per year to the system for a three-year period. This was in addition to such federal expenditures as the states elected to make on urban streets out of primary federal-aid funds, and in some states federal-aid secondary funds. Funds committed under this program during the fiscal year 1955 involved only 329 miles. Iowa, with nearly 26 miles of urban highway improvements programed, accounted for the greatest mileage, but hardly the most important urban highway needs. In each of 28 states less than five miles of urban highways were included in federal-aid projects for which funds were committed during 1955.[30]

In the federal authorization of nearly $1 billion for each of the years 1956 and 1957, money specifically allocated for urban highways totaled $175 million. This was the amount that had to be spent in urban areas, but some states have interpreted the urban area to be the fringe of the city rather than both the city and the urbanized area around it. Moreover, as it is left to the state to determine what cities

[29] The Federal-Aid Road Act of 1916, 39 Stat. 355.

[30] U. S. Bureau of Public Roads, *Annual Report Fiscal Year 1955*, pp. 46–47.

are to receive aid under the federal program, in some states large cities have been neglected despite the intent of the urban provisions of federal legislation. The states also decide how much of the primary federal aid will be spent on urban extensions of the primary system.

In 1939 a report of the Bureau of Public Roads pointed out the urgent need for "the construction of a special system of direct interregional highways, with all necessary connections through and around cities."[31] Subsequently, a committee was appointed by the President in 1941 to recommend a system of national highways designed to meet the requirements of national defense and to provide a basis for improved interregional transportation. In its report to the President and Congress, the committee laid special emphasis on the need for improving a limited system of main highways to high standards and for extension of these routes into and through cities as expressways with full control of access. It was pointed out that "all facts . . . point to the sections of the recommended system within and in the environs of the larger cities and metropolitan areas as at once the most important in traffic service and least adequate in their present state of improvement."[32]

The recommendations were given effect in the Federal-Aid Highway Act of 1944, which called for federal cooperation in aiding the states to designate a National System of Interstate Highways to connect the principal metropolitan areas, cities, and industrial areas.[33] No special funds were authorized for this system, but it was made a part of the primary federal-aid system and was thus eligible, at the option of the states, for improvement with funds for that system.[34] The National System of Interstate Highways was

[31] *Toll Roads and Free Roads,* H. Doc. 272, 76 Cong. 1 sess. (April 27, 1939), p. 4.

[32] *Interregional Highways,* H. Doc. 379, 78 Cong. 2 sess. (January 12, 1944), p. 4.

[33] Federal-Aid Highway Act of 1944, 58 Stat. 838, Sec. 7.

[34] It was not until 1952 that legislation provided funds specifically for the interstate system.

formally designated in 1947. Routes selected at that time totaled 37,700 miles, of which 4,400 were in urban areas. In 1964 additional miles were designated in and around urban areas to bring the system up to the authorized 41,000 miles.

The President's Advisory Committee recommended in 1955 that the federal government assume major responsibility for completion of this interstate system to standards adequate to meet the traffic demands of the next two decades. The cost of the system over a ten-year period was originally estimated at $23 billion, to which $4 billion would have to be added to provide necessary urban connecting arterials. Of the $27 billion total, $15 billion was to be spent in urban areas. The main physical elements of this program were enacted into law in 1956.

Under the terms of the Federal-Aid Highway Act of 1956, the United States embarked on the construction of the National System of Interstate and Defense Highways designed to meet the traffic needs of 1975. A salient feature of the Act was the establishment of the Highway Trust Fund. Under this provision, federal gasoline taxes and other highway-related taxes are reserved in a special highway trust fund which provides grants to the states for road building. In 1961, these funds represented about 44 per cent of the country's highway capital expenditure of $7 billion and 27 per cent of the $11.5 billion spent for all highway purposes, including maintenance, administration, and debt service.[35]

The nationwide highway improvement program launched by the Act of 1956 provides for 90 per cent of the cost of the Interstate System to be defrayed by the federal government and 10 per cent by the states. The system, now estimated to cost $41 billion, will connect 42 state capitals and about 90 per cent of cities with more than 50,000 population. Included are some 6,700 miles of planned access highways in urban areas, including radial arteries, beltways, downtown loops, and bypasses aimed toward decongesting

[35] U. S. Bureau of Public Roads.

central business districts. It is estimated that 45 per cent of the total expenditure will be in urban areas.[36]

As of the end of 1964, a total of 16,963 miles of the interstate system were open to traffic and 6,100 miles were under construction. In many states, more than 50 per cent of the designated system had been completed. (See Appendix Table 12.) The federal program had provided major impetus for an extensive program of modern road construction in metropolitan areas. Thanks to combined federal-state-city efforts, 20 major metropolitan areas at the end of 1964 had completed 2,651 miles of controlled-access urban expressways, and another 1,612 miles were under construction. (See Table 7.)

TABLE 7. Miles of Expressways in Twenty
Major Urbanized Areas, 1964

City	Under Construction	Completed	Total
Atlanta, Ga.	74	50	124
Baltimore, Md.	6	80	86
Boston, Mass.	102	186	288
Buffalo, N.Y.	18	53	71
Chicago, Ill.	96	245	341
Cincinnati, Ohio	98	47	145
Cleveland, Ohio	98	68	166
Dallas, Texas	74	90	164
Detroit, Mich.	59	103	162
Houston, Texas	73	90	163
Jacksonville, Fla.	21	24	45
Los Angeles, Calif.	217	242	459
Miami, Fla.	28	30	58
New York, N.Y.	202	697	899
Philadelphia, Pa.	82	160	242
Pittsburgh, Pa.	51	56	107
St. Louis, Mo.	82	74	156
San Francisco, Calif.	131	183	314
Seattle, Wash.	78	31	119
Washington, D.C.	22	142	164
Total	1,612	2,651	4,263

Source: Automotive Safety Foundation, Urban Freeway Development in Twenty Major Cities, Washington, D.C., 1964.

[36] George M. Smerk, Urban Transportation: The Federal Role, Indiana University Press, 1965, p. 133.

The Larger Network

There are many requirements other than expressways, of course, for adapting America's cities to the automotive age. An extensive system of conventional streets must be maintained and improved, and problems of parking, freight handling and traffic control all add to the complexity and cost of the total task.

There are almost half a million miles of roads and streets within the limits of cities and towns, 12.6 per cent of the total road and street mileage of the United States. A large but undefined mileage of additional highways serves the built-up suburban fringes of municipalities that are part of the urban area but classified as rural. In 1962, some 399,-000 miles of streets within municipal limits were local facilities provided by local governments. In addition to purely local roads, the municipal highway network contained 51,-000 miles of routes that form the urban extensions of the rural state highway systems. Half of this mileage of state highways in cities have also been designated as federal routes eligible for federal aid. The most important of these urban federal-aid highways serve as connecting links for the interstate highway system.

Of the $12 billion spent for highways and streets in the United States in 1963 (exclusive of debt service) $4 billion was spent for streets and highways in incorporated cities and towns. Thus approximately 30 per cent of all money for highways and streets was spent for facilities within the limits of urban places. Part of this money was spent directly by the state for construction and maintenance work done on state and federal routes in the city. Another part was made up of locally expended funds received as grants of motor vehicle tax revenues from the state. These two sources of state aid accounted for 66 per cent of the total outlay on city streets. (See Table 8.) The rest of the money spent for city streets represents funds raised and expended by local governments, obtained from the general tax revenues,

FINANCING OF URBAN HIGHWAYS

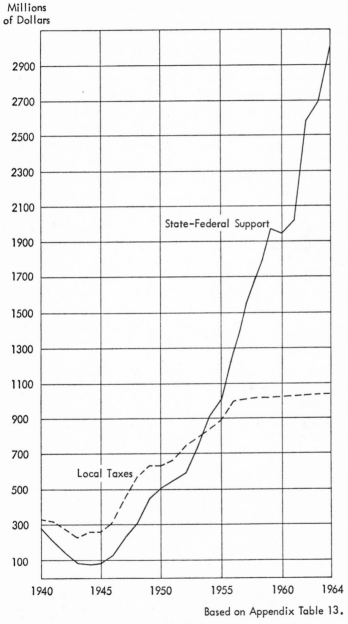

Based on Appendix Table 13.

Chart 8

mainly property taxes but partly from local user charges and tolls.

Increasing state responsibility in urban areas is reflected in the fact that total expenditures for city streets in 1964 were approximately seven times larger than in 1940. An important factor in this increased outlay has been the assumption by the states of responsibility for extensions of the state highway system in cities. The mileage of these urban links was four times greater in 1964 than in 1934, and the amount of state effort on these routes had substantially shifted the financial burden. Expenditures by the state for urban extensions of the state system (including federal grants spent through the states) rose from $336 million in 1950 to $2.5 billion in 1964. At the same time cash grants from state user revenues to cities for expenditures on local streets by the municipal highway departments increased from $175 million in 1950 to $525 million in 1964. (See Chart 8.) Thus, the combined amount of state aid trans-

TABLE 8. Highway Expenditures in Cities, 1963

	Expenditures (millions)
State highway department expenditures on urban extensions of the state system	$2,190
State aid to cities for streets	503
Municipal street expenditures from local sources	1,401
Total	$4,094

Source: U. S. Bureau of Roads, Tables HF-2, SF-2, and SF-4.

ferred to municipalities for city streets plus direct expenditures by the state highway department on urban extensions of the state system yielded total state aid of nearly $3 billion in 1964.

The fact remains, however, that an adequate highway improvement program has not been possible in many of the major cities because rurally dominated state legislative bodies and state highway departments have had limited jurisdiction and limited interest in the big city. As a consequence, the metropolis which generates large amounts of state-collected motor vehicle tax revenues has seen a disproportionately small share of these revenues spent for the facilities urgently needed to combat urban congestion.

The states established their highway departments at a time when the problem of providing rural highways was the major task, and at the beginning these organizations were restricted by law in their contributions to the solution of urban problems. State-collected motor vehicle and gasoline tax revenues were dedicated to work on rural roads. State activity stopped at the municipal boundary, where generally some type of surfacing had been provided by the city at an early date.

Parking and Terminal Facilities

Closely related to street requirements is the need to get the parked vehicle off the street to provide room for moving traffic. Off-street parking facilities, both above ground and underground, are gradually replacing curb parking. In 1962, gross private and public investment in parking facilities was more than $5.25 billion. Private operations account for the bulk of the investment, but public parking has played a significant role. Capital costs of public facilities have been financed both with city tax funds and through municipal bond issues, and to a considerable extent these facilities have been self-supporting.[37]

Many local units of government have zoning ordinances requiring off-street parking accommodations in all new or substantially remodeled buildings. The ratio of parking space to floor area varies with contemplated uses. In Chicago, parking and loading requirements incorporated in the zoning ordinance pertain to all uses generating automobile

[37] Committee for Economic Development, *Developing Metropolitan Transportation Policies: A Guide for Local Leadership,* 1965, p. 23.

and truck traffic. The law applies to all of the city except the central business district. There, parking facilities are not compulsory because of extremely high land values and the feasibility of providing public parking in garages and lots.[38]

In many central cities downtown off-street parking has been provided through various combinations of public and private approaches. One method is illustrated by a garage development program in Pittsburgh. Since 1955, nearly 6,000 spaces have been provided in eight major downtown garages. Six of them have been constructed by the Public Parking Authority, established in 1947, and three have been leased by major department stores. These facilities are serving store patrons as well as commuters working at new office buildings.

An underground garage was financed and developed by a private concern as part of the Gateway Center complex of hotels and office towers. In addition, over 1,000 surface parking spaces have been provided by the Auditorium Authority near the Civic Arena for use by downtown employees during the day and by spectators attending special events at night.

In San Francisco the Parking Authority of the City and County was established in 1949, with the aim of encouraging private development of off-street parking. Accordingly, in recent years about two thirds of all downtown parking spaces have been provided solely by private capital, and the remainder through public-private co-operation. Private enterprise has provided nearly 23,000 spaces downtown since 1949 and as of 1962 the city's intensely developed central business district contained nearly 12,000 spaces in 33 private and public garages. The city also acquires land through condemnation; then a non-profit parking corporation builds a garage and assumes only the cost of the garage (not the land). After the original investment is amortized, the facility reverts to municipal ownership. A

[38] Evert Kincaid, "Chicago's New Off-Street Parking and Loading Ordinance," *Traffic Quarterly* (April 1954), p. 246.

number of underground garages have been constructed through this procedure.[39]

Perimeter parking is also provided in many cities to help reduce the volume of traffic in downtown areas. These lots are located outside high density areas and adjacent to expressways and transit lines leading to the center. Such arrangements have had different degrees of success, but where location and connecting public carrier services have met with the needs of commuters, they have attracted new customers to transit and have reduced the volume of automobile traffic that would otherwise have entered the downtown area.

Until recently only minor attention was given to the provision of adequate truck and bus terminals. The truck problem and its impact on passenger traffic has grown to serious proportions. A study in Milwaukee found that on weekdays trucks accounted for one fourth of all trips in the metropolitan area. In view of the fact that much of the passenger car travel was on boulevards and parkways, traffic on the average city streets was estimated to be more than one third commercial traffic.[40] Normally, each of the truck lines serving an urban area picks up and delivers all the freight it carries. This means that each route may be duplicated by several truckers. The volume of trucking on city streets, therefore, is much greater than would be necessary if some consolidation of shipments were possible.

Properly located and designed union terminals not only reduce street congestion but distribution time and costs as well. Small trucks can pick up freight in specified areas for delivery to the terminal where it is sorted and consolidated for the over-the-road operation. Likewise large trucks arriving at the terminal can have their loads sorted and transferred to smaller delivery vehicles, each destined for a particular area of the city.

[39] Wilbur Smith and Associates, *Parking in the City Center,* New Haven, 1965, pp. 53–55.

[40] William R. McConochie, "Trucks, Traffic and Terminals," *Traffic Quarterly* (October 1949), pp. 329–33.

More efficient organization of freight movement is important to the city, the trucking industry, and the shipper and consumer. In New York City it was concluded that seven to ten times as many trucks use the streets as would be necessary if consolidation of shipments were possible. With the opening of the Manhattan Union Terminal, it was estimated that one truck was able to perform services that required 2.25 vehicles prior to the consolidation.[41] A study of the truck terminal problem in Boston indicated that truck traffic could be cut in half by the provision of a union terminal, plus substantial savings in cargo handling and truck operation.[42]

The scattered and indiscriminate location of more than 200 truck terminals in Chicago has been a primary factor in traffic congestion in the Loop. It was estimated that 5,000 trucks were parked on city streets every day. At one time the amount of trucking required from one terminal to another involved 15,000 to 20,000 truck trips daily, carrying more than one fourth of all truck tonnage moving in the city. On an average business day, 13,000 trucks were entering the central business district and making 26,500 stops. The average amount of merchandise picked up or delivered was less than half a truck load. Deliveries to one department store in the course of a day involved 378 trucks, but store officials estimated that all this freight could have been delivered in consolidated loads using 13 vehicles.[43]

A survey of truck transportation patterns in Chicago led to the conclusion that the establishment of truck terminals along the outer perimeter of the city would be impractical. Concentrating them near each other, and close to most of

[41] D. L. Sutherland, "Union Truck Terminals," *1950 Proceedings,* Twenty-First Annual Meeting, Institute of Traffic Engineers (1950), p. 24.

[42] Boston City Planning Board, *Report on a Union Motor Truck Terminal for Boston* (October 1947).

[43] Phil Hirsch, "Chicago Fleetmen Battle Traffic Bottlenecks," *Commercial Car Journal* (June 1954), pp. 68–69, 178–86.

their delivery points, would do the most to reduce traffic among terminals and to consolidate shipments. A new zoning classification known as Truck Terminal Districts was created by the city and four such zones were established to house most of the trucking companies of the city. It was stipulated that the area devoted to these zones should be no less than 25 acres and that location should be close to planned expressway routes. The first of four planned terminal zones consisted of 20 square blocks designed to house 40 truck terminals serving 60 major truck lines. The four zones were expected to divert more than 12,000 trucks or about 50 per cent of all heavy duty traffic from Chicago streets.[44]

Essentially the same reasons that have led to the provision of union truck terminals call for the establishment of terminals for public carriers of passengers, such as the Port of New York Authority Bus Terminal. In May 1963, the Greyhound Corporation consolidated all of its New York City operations in the newly expanded bus terminal. With this move, the terminal has been able to house the country's two major long-distance carriers, Greyhound and Trailways. Total traffic at the terminal during 1963 was 63 million passengers on 25 million buses.[45] The diversion of traffic on overhead ramps from the terminal to the Lincoln Tunnel has been equivalent to adding three cross-town streets. As much as 30 minutes per bus trip is saved by using the off-street facility. In addition, the Port Authority building provides close to 200 truck berths for the consolidation and distribution of truck and rail freight. Thus the terminal provides a considerable degree of traffic relief by keeping both trucks and buses off the streets.

[44] *Fleet Owner* (September 1952), p. 63. Relocation of all 495 trucking companies into these and other areas yet to be developed will involve an investment by trucking companies in excess of $100 million. Over a 20-year period it is anticipated that savings in operating costs will more than compensate the trucking industry for its investment.

[45] Port of New York Authority, *Annual Report* 1963, p. 2.

The Port Authority Bus Terminal has not only provided improved service for the commuter and relief of traffic congestion on the surface streets, but has afforded an opportunity to develop concessions that have made an important contribution to terminal revenues. There are more than sixty commercial and manufacturing tenants in the building.

The Continuing Dilemma

Progress in the construction of highways and terminals has been notable, but the task of adapting to the automotive age is much more than the function of supplying transport capacity. Efforts to make up for lost time through multi-billion dollar transportation programs can be nullified by the expansion of population and economic activity and by the character of new urban development. Cities and their suburbs are in the midst of a construction boom that often gives little recognition to the automotive environment. Many new office buildings are being constructed where congestion has already reached critical proportions. These buildings generally house far more people than the smaller structures they replace. In most cases no provision is being made, or can be made, to augment the transportation facilities that serve these new traffic generators. Many of these structures fail to include adequate off-street parking and truck loading and unloading facilities.

The whole pattern of urban development today tends to ignore how people move, and how they will be moving in the decades ahead. Building heights, densities of population, and the amount of ground being covered by new development are dooming costly expressway programs everywhere. In many urban areas the traffic problem seems to be worsening much more rapidly than new highways can furnish relief.

In the suburbs the problem is no less critical. Failure to preserve open space to balance new urban growth subjects the fringes to the same problems of traffic congestion that suburbanites were seeking to avoid in the first place. The automobile has expanded our radius of travel and permitted the urban resident a fuller life. But the physical deterioration of the community, the lack of amenities, and the disappearance of open space means that the urban resident must drive farther and farther to reach his destination. We are engaged in a race between the increasing mobility provided by the automobile and the highway, and the increasing distances that have to be traveled from home to work, from home to recreation, and from city to country. Unless the development of a better urban environment accompanies the development of better highways, the race will be lost.

In the midst of an extraordinary effort to provide extensive new highways, therefore, doubts have arisen in many communities regarding the ultimate effectiveness of the current approach. Traffic congestion in urban areas and the outlook for continued growth of population and automobile ownership raise the question whether the big city is capable of adapting to the automobile. The engineering difficulties and heavy financial outlays involved raise the further question whether the attempt to adapt is worth the cost. Thus far few urban areas have accomplished anything approaching the system of expressways and terminals needed to accommodate the dense traffic of central cities and the continuing expansion of the area of urbanization. And few cities have adequately planned their future development to know where these highways ought to be located and how they might serve to promote the realization of better communities. Under these conditions the possibilities of adapting to the automotive age are not reassuring.

CHAPTER III

The Crisis in Public Transportation

Hope of meeting the transportation needs of the urban area through greater reliance on public transportation have been dimmed as the fortunes of the industry have continued to decline. The cycle of traffic losses, rising costs, higher fares, less frequent service, and further loss of business has persisted. The same growth factors that created the overwhelming demand for automotive transportation have caused a sharp reduction in the patronage of mass transportation facilities. Prosperity for the motor age has meant depression for transit.

These conditions, added to the difficulties of driving, might have been expected to encourage more energetic efforts to improve public transit. But the policy of the cities has been negative. The fiction has been perpetuated that the provision of transit services is a privilege and a monopoly, and that the role of government is to protect the public against financial abuse. Generally, no public responsibility for mass transportation has been assumed until financial crises have demanded intervention. Transit has been divorced from other sectors of urban transportation, despite the fact that policies governing these other sectors have a vital impact on the transit system.

The current situation stems from a combination of historical factors, characteristics inherent in the industry, and public policies subject to revision. The present chapter will review the economic behavior of the industry and the policies that affect it in order to provide a basis for appraising the future of mass transportation.

An estimated 16 to 18 million families depend on public transportation by street car, bus, and subway or elevated lines. Measured by annual revenues, the transit industry is the largest public passenger carrier engaged in ground transport. Transit revenues are nearly twice the amount collected from railroad passenger service and three times the revenues collected by the nation's intercity bus operations.

From the standpoint of traffic volume and revenues, the transit problem is localized in a relatively small number of large cities. The ten largest cities accounted for almost half of all mass transportation revenue passengers in 1963, and the 41 largest accounted for 71 per cent of the passenger load. The largest operation was in New York City, where the transit system in 1963 accounted for revenues of $288 million, representing 21 per cent of all transit revenues in the United States and 26 per cent of total traffic. Chicago transit was second largest with revenues of $138 million. Other cities with revenues of over $25 million included Philadelphia, Newark, Los Angeles, Boston, Washington, D.C., Detroit, and Cleveland.[1]

The importance of the role of mass transportation varies among cities of different size, type, geographical location, and income. The larger the city the higher the proportion of transit travel into and out of the downtown areas. In a number of cities one half to two thirds of all persons entering the central business district do so by transit. They include such places as Chicago, New York, Newark, Philadelphia, Richmond, Boston, Atlanta, and Cleveland. In these cities the physical layout had crystallized long before the availability of the motor vehicle, and resulting high population densities have made a substantial dependence on mass movement essential. Other large cities, such as Dallas and Los Angeles, which grew in a motor age, are less intensively developed and more dependent on the automobile. (See Chart 9.) So are smaller cities, where concentration of population is not so great, and where the problems of automobile traffic and parking are less severe.

[1] Data supplied by the American Transit Association.

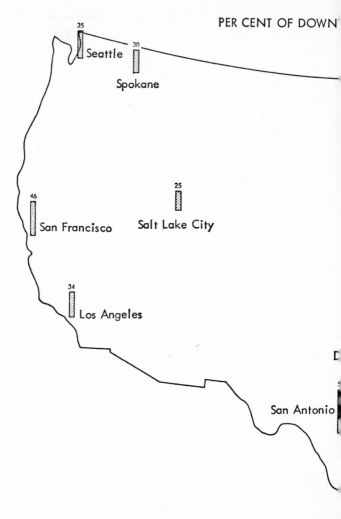

35 Seattle

30 Spokane

46 San Francisco

25 Salt Lake City

34 Los Angeles

San Antonio

Chart 9

L BY TRANSIT

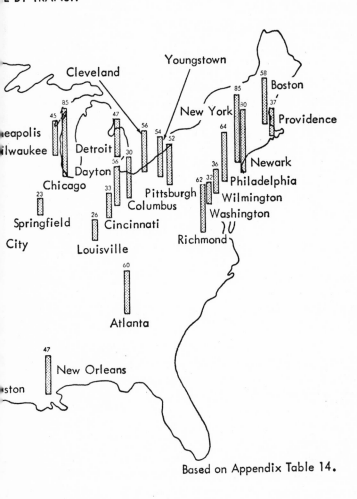

Based on Appendix Table 14.

Trends in Mass Transportation

The trend away from public carriers is indicated by figures of consumer expenditures. Consumers in 1963 were spending $47 billion per year for transportation, of which $43.5 billion was spent for automobiles and their operation. This outlay for automobiles compares with $2 billion spent in cities for local public carriers. Expressed in 1954 dollars, the increase in expenditures for local public carriers was from $1.8 billion in 1941 to $2.2 billion in 1954, whereas private automobile expenditures rose from $13.2 billion in 1941 to $23.7 billion in 1954. Between 1956 and 1963, however, expenditures for local public carriers declined from $2 billion to $1.8 billion, while automobile expenditures kept rising from $30 billion to $38 billion. (See Chart 10.)

In 1909 local public carriers commanded 33 cents out of every dollar spent by the consumer for transportation. By 1929 this proportion had fallen to 15 cents and in 1963 only four cents out of every transportation dollar was going for transit. Over an extended period, therefore, the proportion of transport expenditures devoted to local public transportation has been declining, so that the belief that improvement of mass transportation can effect a sizable shift from the automobile assumes a reversal of consumer attitudes that have prevailed over a considerable period of time. (See Table 9.)

Transit rides purchased in 1964 totaled only 37 per capita compared with 132 in 1940 and 162 in 1954. The total volume of transit riding from 1940 to 1962 declined 34 per cent despite the fact that gross national product increased 139 per cent and industrial employment 68 per cent. (See Chart 11.) From 1930 to 1940 the decline in transit riding was 16 per cent. Then came the war, and with it gasoline rationing and the end of automobile production. A great increase in public transportation occurred in all but the largest cities. The increase for the country as a whole was 32 per cent. In cities of a million people and over the transit

CONSUMER EXPENDITURES FOR AUTOMOBILE TRANSPORTATION AND LOCAL PUBLIC CARRIERS

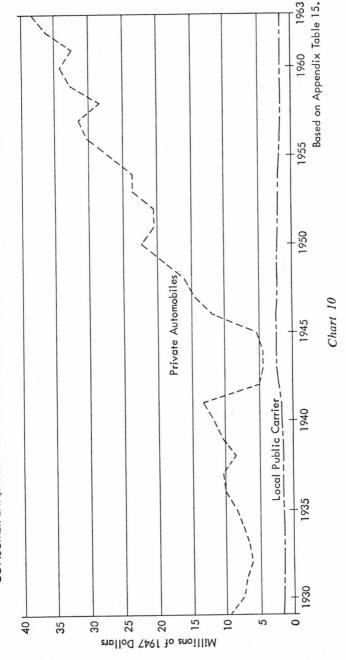

Chart 10

Based on Appendix Table 15.

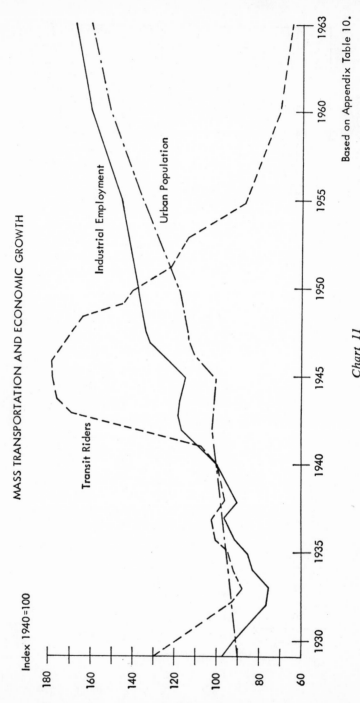

MASS TRANSPORTATION AND ECONOMIC GROWTH

Index 1940=100

Transit Riders

Industrial Employment

Urban Population

1930 1935 1940 1945 1950 1955 1960 1963

Based on Appendix Table 10.

Chart 11

TABLE 9. Percentage Distribution of Consumer
Expenditures for Transportation[a]

Type of Transportation	1909	1919	1929	1940	1950	1954	1963
All Transportation	100.0	100.0	100.0	100.0	100.0	100.0	100.0
Private	47.4	71.5	78.3	82.2	86.9	87.9	92.1
Local public carrier	33.3	16.8	14.7	12.7	8.9	8.0	4.2
Intercity public carrier	19.3	11.7	7.0	5.1	4.2	4.1	3.7

[a] J. Frederic Dewhurst and Associates, *America's Needs and Resources,* Twentieth Century Fund (1955), p. 971; U. S. Department of Commerce, *National Income,* A Supplement to the Survey of Current Business (1954), pp. 206–7; *Survey of Current Business* (July 1955), p. 19; *Survey of Current Business* (July 1964), p. 16.

boom added less than 5 per cent to the total number of passengers. In the smaller cities and the new suburbs, however, the increase in transit riders was tremendous. Cities of 50,000 population and less experienced a 150 per cent increase in transit business. There was a 95 per cent increase in cities with 50,000 to 100,000 people.

The sharp rise in transit patronage during the war years was naturally followed by a precipitous drop as the automobile was permitted to resume its role. But that drop has continued in the 1950's and 60's. (See Table 10.) Between 1950 and 1955 the decline was 33 per cent for the country as a whole, and by 1963 there was a further decrease of 27 per cent. Cities of all sizes have contributed to the de-

TABLE 10. Percentage Change in Transit Patronage
by Size of City

Population Group	1930–40	1940–50	1950–55	1955–63[a]
1,000,000 and over	− 9.6	4.7	−26.0 ⎫	−13.3
500,000–1,000,000	−17.8	39.0	−35.0 ⎭	
250,000–500,000	−10.3	46.3	−35.8	−55.1
100,000–250,000	−15.9	53.0	−39.8	−53.5
50,000–100,000	−11.5	95.3	−39.9	−38.1
Less than 50,000	9.0	150.1	−49.1	−42.3
Total	−15.9	31.7	−33.2	−27.1

Source: Data supplied by the American Transit Association.

[a] American Transit Association, *Transit Fact Book, 1956,* p. 5; *Transit Fact Book, 1964,* p. 6.

cline. The net impact of the decline was least in the smaller
cities where the transit boom of the previous decade had
been so spectacular. But in the largest cities, where transit
had been relatively unaffected by the impacts of the war,
there was a 26 per cent drop in transit riding between 1950
and 1955, and a further decline of 13 per cent between
1955 and 1963. The greatest decline has thus occurred
where the problems of traffic congestion are the worst.

To illustrate, Chicago's transit riders in 1963 numbered
only 58 per cent of the 1940 total, while the drop was to 38
per cent in Pittsburgh, 43 per cent in Detroit, and 62 per
cent in Los Angeles. (See Table 11 below.)

TABLE 11. Index of Transit Revenue Passengers in
Major Cities, 1963[a] (1940 = 100)

City	Passenger Traffic Index
New Orleans, La.	114.7
San Antonio, Texas	98.5
San Francisco, Calif.	81.2[b]
New York, N.Y.	76.2
Manhattan & Bronx STOA	74.9[c]
Buffalo, N.Y.	74.3
Boston, Mass.	67.9[d]
Seattle, Wash.	67.4
Washington, D.C.	67.3
San Diego, Calif.	67.2
Dallas, Texas	67.0
Baltimore, Md.	64.2
Philadelphia, Pa.	64.0[e]
Los Angeles, Calif.	62.1
Cincinnati, Ohio	58.0
Chicago, Ill.	57.7
Milwaukee, Wis.	55.1
St. Louis, Mo.	47.3[f]
Detroit, Mich.	42.8
Pittsburgh, Pa.	37.7

Source: Data supplied by the American Transit Association.

[a] Cities with population in excess of 500,000.
[b] 1940 figures exclude California Street Cable Railroad Co., which was absorbed by Municipal Railway in 1952.
[c] Formerly Surface Transit Inc. and Fifth Avenue Coach Lines, New York, New York.
[d] 1960, the latest available figure.
[e] 19-day strike, January–February 1963.
[f] 1962.

Trends in Mass Transport Methods

The decline in transit riding has been accompanied by a rapid transformation from street car to rubber-tired transportation. Between 1917 and 1928, surface street car lines were carrying 12 to 13 billion passengers annually. In 1963 only 300,000 passengers were transported by street car. The bus has been able to meet the new requirements of route flexibility imposed by the widely dispersed origins and destinations of traffic resulting from growth of the metropolis in an automotive environment. The inflexible nature of rail transportation, on the other hand, has prevented an adjustment to the crisscross patterns of the new mobility. Rapid transit operations have remained at a fairly even level for half a century, with a slow decline since the war to levels prevailing in the early 1920's. But the percentage of total patronage has increased from 15 per cent in 1954 to 22 per cent in 1963. (See Chart 12.)

The fortunes of the street car and rapid transit line have been partially duplicated by trends in railroad commuter transportation. Since 1929 the number of commuters traveling to and from cities by railroad has declined more than 45 per cent. This trend took place during the prewar decade, and although an upsurge was recorded during the war, total traffic volume has again returned to prewar levels. There were only 195 million commutation tickets sold in 1963 compared with 459 million in 1929.[2]

The shift from public carriers to private automobile and the shift in the public sector of the transport system from rail to bus have meant a sharp increase in the space de-

---

[2] Shifts from rail to rubber have occurred in the movement of freight as well as passenger transportation. The number of trucks in the United States in 1964 exceeded 13 million, and the proportion of total freight movements into and through the urban area by rail had fallen from 61 per cent in 1940 to 44 per cent in 1964.

mands of transportation in urban areas. It is these shifts, combined with population and economic growth, that have accentuated the inadequacy of outmoded street systems and multiplied the demand for new highway and terminal capacity.

SHIFT IN METHODS OF MASS TRANSPORTATION

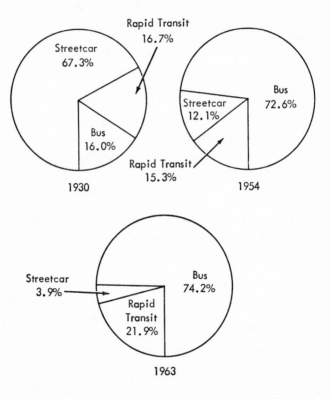

Based on Appendix Table 16.

Chart 12

This shift has meant that the vehicular composition of big city traffic patterns has become heavily weighted by automobile operation. A study conducted in Chicago in 1954, for example, showed that 70 per cent of the vehicles using the streets were private automobiles. It should be noted, however, that these vehicles carried only 27 per cent of passengers entering the central business district. Despite the trend toward automobile use, the public transportation system still carried the major load in this close-in area. Rapid transit by subway and elevated, which accounted for about the same number of passengers as the private automobile, operated fewer than 3 per cent of the vehicles entering the central business district. None of these vehicles operated on the streets. Public carriers on the streets, including electric cars and buses, carried over 20 per cent of total passengers with another 3 per cent of the vehicles. (See Table 12.)

TABLE 12. Passenger Transportation Pattern, Chicago
Central Business District, 1954[a]

Type of Vehicle	Percentage of All Vehicles	Percentage of All Passengers
Street cars	0.60	6.58
Local buses	2.37	14.02
Out-of-town buses	0.33	1.15
Private autos	70.21	27.47
Service vehicles	9.90	2.58
Taxicabs	12.99	5.08
Subway and elevated cars	2.45	27.57
Railroad cars	1.15	15.55
Total	100.00	100.00

[a] Data from cordon counts made principally by Chicago Bureau of Street Traffic and co-ordinated by Chicago Association of Commerce and Industry. Werner W. Schroeder, *Metropolitan Transit Research,* Chap. 2 (December 20, 1954), Addendum C, Tables 1 and 2.

New York City commuter trends have paralleled in major respects the experience of Chicago. The number of passengers crossing the Hudson between New York and New Jersey by all forms of transportation has risen very little over the past three decades. In 1930 trans-Hudson passengers numbered 262 million. The 1954 figure was 281 million, after which there was an increase to 314 million in 1961. But this higher figure was still only 20 per cent above the 1930 level.

But the most significant changes have taken place in the
methods of trans-Hudson movement. Railroads lost 117
million passengers from 1930 to 1961. The number of
pedestrians using ferries (other than those operated by the
railroads) declined from 38.6 million in 1930 to a mere
0.4 million in 1961. Buses (other than rail-operated) increased their patronage by nearly 76 million passengers in
the same period, and automobile riders increased by 13
million. Auto and bus together accounted for 207 million
more riders per year than in 1930, whereas railroad and
ferry passengers in 1961 were 155 million fewer than in
1930.[3] On a percentage basis, railroad patronage was down
68 per cent between 1930 and 1961; pedestrians via ferry
had declined 99 per cent; bus passengers were up 876 per
cent; and auto riders had increased 300 per cent. (See
Table 13.)

TABLE 13. Trend in Cross-Hudson Passenger Movement

Mode of Trans-portation	1930		1954		1961		Percent-age Change 1930–61
	Millions of Pas-sengers	Per Cent	Millions of Pas-sengers	Per Cent	Millions of Pas-sengers	Per Cent	
Rail (by ferry, tunnel, and railroad bus)	170.7	65.2	75.3	26.8	53.8	17.1	−68
Bus (excluding railroad bus)	8.7	3.3	67.5	24.0	84.9	27.0	876
Pedestrians via ferry	38.6	16.7	2.8	1.0	0.4	0.1	−99
Automobile	43.7	14.8	135.3	48.2	175.0	55.7	300
Total	261.7	100.0	280.9	100.0	314.1	100.0	20

Data from tabulations of Planning Division, Port of New York Authority,
dated March 1955.

Source: Comprehensive Planning Office, the Port of New York Authority,
Metropolitan Transportation—1980, 1963, p. 340.

For the New York City area as a whole, the peak in
suburban rail commuting occurred in 1929–30, and a de-

[3] Data from tabulations of the Port of New York Authority.

cade and a half later, despite the great increase in the suburban population of New York, there were 48 per cent
fewer rail commuter trips than at the peak. From 1950 to
1961 there was a further decline of 29 per cent.[4] Employment in manufacturing and wholesale trade has been growing rapidly in the suburbs, contributing to a sharp decline
in the number of commuters to New York per thousand
suburban families.[5]

Peak-Hour Problem

The decline of mass transportation fails to reflect the
continuing importance of public carriers in the rush hour.
This peak-hour role is indicated by a survey of trips leaving
the Philadelphia central business district during a typical
weekday in 1955. (See Chart 13.) For the entire period of
24 hours, mass transportation accounted for 50 per cent of
the total trips leaving the area. During the 5 to 6 p.m. peak
hour, however, nearly 72 per cent of all trips from the
center were by public carriers. Conversely, during the entire 24 hours the automobile accounted for 41 per cent of
persons leaving the business district, whereas during the
peak hour only 24 per cent of total trips were in automobiles or taxis.

The special importance of rail service in Philadelphia
during the rush hour is demonstrated by the fact that
whereas movement by rapid transit and suburban railroad
accounted for 27.4 per cent of total trips leaving the Philadelphia area at all times of the day, they comprised 47.7
per cent of all trips during the 5 to 6 p.m. rush. The problems for the industry that result from this concentration of

[4] Comprehensive Planning Office, the Port of New York Authority, *Metropolitan Transportation—1980,* 1963, p. 292.

[5] Regional Plan Association, Inc., *New York's Commuters,*
Bulletin 77 (July 1951).

MOVEMENT OF PEOPLE FROM DOWNTOWN PHILADELPHIA

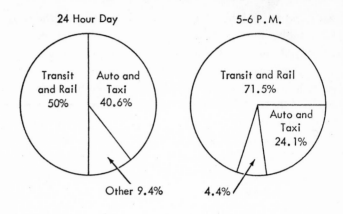

24 Hour Day

Transit and Rail 50%

Auto and Taxi 40.6%

Other 9.4%

5-6 P.M.

Transit and Rail 71.5%

Auto and Taxi 24.1%

4.4%

Per Cent of Each Carrier's Daily Traffic in the 5-6 P.M. Rush

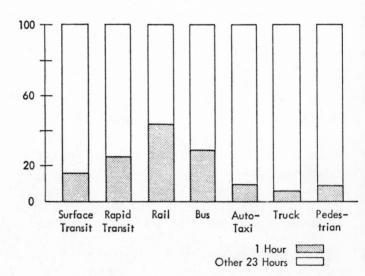

Surface Transit · Rapid Transit · Rail · Bus · Auto-Taxi · Truck · Pedestrian

1 Hour
Other 23 Hours

Based on Appendix Table 17.

Chart 13

public carrier business are apparent. In one hour the commuter railroads performed 44.5 per cent of their total outgoing business for the day. Rapid transit carried 25.5 per
cent of the day's outgoing travel in this one hour. On the
other hand, the automobile and taxi performed only 9.7
per cent of their daily outbound movement during the 5
to 6 p.m. rush hour.

Travel in the New York metropolitan area demonstrates
this tendency to rely on mass carriers primarily during a
few hours of the day when traffic volumes in or related to
the central business district restrict the movement of private
automobiles. (See Table 14.) On a typical business day in
1960, the 24-hour count of passengers entering downtown
Manhattan revealed that 26 per cent were moving by automobile. Mass transportation carriers were accommodating
70 per cent. In the morning rush hours, from 7 to 10 a.m.,
however, the automobile was carrying only 13 per cent of
the load and public transportation accounted for 85 per
cent.

These facts about the rush hour explain why the downward trend in transit riding since the war has been so much
more destructive to the industry than total traffic figures
indicate. The mass transportation problem has been magnified by the fact that traffic has continued high during
peak hours of the day, while most of the loss of business
has been in off-peak hours. This pattern of passenger movement means that the transit company must still meet the
high man power and equipment requirements dictated by
the needs encountered in the peak hours—generally two
hours in the morning and two hours at the end of the working day—despite the over-all reduction in number of passengers accommodated throughout the day. Much of this
equipment is idle during the off-peak period, and the working force that must be employed for a few hours to handle
the peak must be paid for a full day.

This pattern of transit traffic poses the double threat of

TABLE 14. Passengers Entering Manhattan's Central Business District on Typical Business Day, 1960
(In thousands)

Mode of Transportation	Total Day (24 hours)		Peak Hours (7–10 a.m.)	
	Passenger Volume	Percentage of Total Passengers	Passenger Volume	Percentage of Total Passengers
Rapid Transit	1,913	57.1	1,133	69.6
Commuter Railroad	203	6.1	143	8.8
Bus[a]	243	7.3	101	6.2
Ferry[b]	36	1.1	25	1.5
Auto and taxi	866	25.9	204	12.5
Other[c]	88	2.6	21	1.3
Total	3,349	100.0	1,627	100.0

Source: Comprehensive Planning Office, the Port of New York Authority, *Metropolitan Transportation—1980*, New York, 1963, p. 295.

[a] Excludes rail passengers delivered to Manhattan by bus.
[b] Pedestrian passengers only.
[c] Passengers who entered driving or riding in trucks.

bankruptcy for the industry and prolonged deterioration of service for the customer. For the decline in transit patronage has not permitted parallel reductions in cost, and emergency measures designed to bring costs more in line with revenues have resulted only in reducing the attractiveness of the service. The effect has been to create an even greater incentive for car owners to desert rail and bus transportation for their automobiles.

The slump in off-peak traffic and the maintenance of a high peak-hour demand can be explained by the heavy congestion on city streets during rush hours and the lack of all-day parking space. This has made it necessary in many cities to use mass transportation to get to work; and the choice has been a logical one for the commuter because a large proportion of home-to-work trips are along established transit routes converging in the downtown area. But at other times of the day and night, the automobile is preferred to mass transit as less congested traffic makes driving tolerable. It is then that the even rhythm of peak-hour commutation movement between suburb and city changes to a discord of heterogeneous trips in all directions for social and recreational purposes.

To illustrate the problem, in the Boston metropolitan area the number of transit riders decreased 38 per cent from 1946 to 1953. But the decline in patronage during rush hours was only 10 per cent. Thus for a period of about four hours every weekday the equipment and man power requirements of the transit system were little less than they were in 1946, while traffic and income for the system as a whole was sharply down.[6] In New York the number of people using the subways in the 7 to 9 a.m. and 5 to 6 p.m. rush hours had fallen only slightly since the war. Rush-hour traffic volume in 1951 was 96 per cent of 1945, but

[6] Warren H. Deem, *The Problem of Boston's Metropolitan Transit Authority,* Publication No. 20, Bureau for Research in Municipal Government, Graduate School of Public Administration, Harvard University (1953), p. 41.

traffic during the evening was only 70 per cent of 1945 and night traffic had declined to 72 per cent. (See Table 15.)

TABLE 15. Average Daily Distribution of New York Subway Passengers by Time of Day, 1945 and 1951[a]

Time of Day	Thousands of Passengers				1951 as Percentage of 1945
	1945	Per Cent	1951	Per Cent	
Rush hours 7–9, 5–6	2,244	36.8	2,156	39.4	96.1
Near rush 6–7, 9–10, 4–5, 6–7	1,443	23.6	1,307	23.9	90.6
Midday 10–4	1,278	20.9	1,199	21.9	93.8
Evenings 7–12	892	14.6	626	11.5	70.2
Night 12–6	247	4.1	178	3.3	71.9
Total (24 hours)	6,104	100.0	5,466	100.0	89.5

[a] William S. Vickery, *The Revision of the Rapid Transit Fare Structure of the City of New York,* The Mayor's Committee on Management Survey of the City of New York, Technical Monograph No. 3 (February 1952), p. 86.

The cost of accommodating the demand for public transportation is indicated by data from the Metropolitan Transit Authority in Boston. Off-peak requirements were so much lower than the peak that only about one third of the buses could be used all day. They carried 53 per cent of total traffic. Another third of the fleet went into service with the approach of the rush hours and came out after the rush and carried about 33 per cent of total passengers. The final third operated for only the four peak hours and these buses accounted for only 14 per cent of total riders. The buses operated throughout the day or over the entire rush-hour period returned almost four times as much revenue as the buses operated only during the peak.[7]

In addition to off-peak hours, public transportation must also cope with off-peak days. The five-day work week has contributed significantly to the downward trend of transit patronage since the war. In the Boston metropolitan area in 1946, the number of Saturday transit riders was 93 per cent of the average number of weekday riders; and on

[7] Ralston B. Smyth, "The High Vehicle Cost of Peak Hour Bus Service," *Mass Transportation* (February 1953), p. 23.

Sunday the transit system carried 52 per cent of the week-day load. By 1952, however, the five-day week had reduced Saturday transit patronage to only 74 per cent of the week-day average, and on Sunday the load fell to 35 per cent of weekday travel.[8] Trends in the Chicago transit pattern have been much the same. In 1940 the volume of Sunday riders on Chicago surface lines was between 85 and 90 per cent of average weekday traffic. During 1953, however, Sunday riding on surface transit vehicles ranged from 42 to 50 per cent of the weekday average, and rapid transit lines were carrying only 26 to 30 per cent of weekday loads.[9]

The peak-load problem is not peculiar to transit, but the peak for the transit industry is more severe than that of other utilities such as electric, gas, and telephone companies, and differs in that the latter enjoy a degree of monopoly that permits rates to be adjusted to meet costs. The transit industry, in competition with the private automobile, does not have a free hand in adjusting rates to traffic conditions. Moreover, the telephone company can charge according to type of service, time of day, and length and type of call, whereas transit fares tailored to reflect distance, type of service, class of equipment, and time of day introduce much greater complexities. The financial impact on the industry from peaks and valleys in transit riding cannot be compensated, as in the case of railroad passenger service, from freight revenues that absorb the passenger deficit.

Along with unprofitable hours of work, the transit industry must also cope with unprofitable routes. Such routes have increased in mileage with rising costs, lower densities of population encountered in newly developed suburban areas, and increasing numbers of two-car families living in the suburbs. Most transit companies use a large part of the earnings from profitable lines to underwrite losses on poorly paying lines. In Providence, Rhode Island, for ex-

[8] Edward Dana, "Statement before the Committee on Metropolitan Affairs of the Massachusetts Legislature" (January 19, 1953).

[9] Schroeder, *Metropolitan Transit Research,* Chap. 2 (December 20, 1954), p. 27.

ample, 22 of the 38 lines of the company showed a loss. The remaining 16 lines produced 57 per cent of the gross earnings of the entire system. The poorly paying lines failed in most cases to come close to the 54 cents a mile that was the average cost of operation for the entire system. Profitable lines earned 67 cents per mile.[10] In Kansas City, Missouri, 24 of the 43 lines operated by the transit company failed to pay their way.[11]

Public pressure to maintain unprofitable routes is strong, however, and the difficulties of obtaining the consent of regulatory bodies to abandon service are considerable. The question whether the service should be maintained regardless of its profitability creates a dilemma for public utility commissions. From the standpoint of the community, it may be that the service is beneficial. But from the standpoint of the transit industry, it is uneconomical and in direct opposition to the need of the company to keep its operations in the black.

Rising Costs and Fares

While suffering a decline in passengers, the transit industry has been experiencing a steady rise in costs of operation. All major companies have had to meet sizable wage increases in the inflationary period since the war. The cost of transit labor has increased over 100 per cent in many cities, and the bill for labor ranges from 45 to 80 per cent of the total annual operating costs of the industry. Between 1947 and 1963 there have been 811 costly transit strikes. Fringe benefits have added 10 to 20 per cent to labor costs. Replacement parts and fuel costs are about double what they were before World War II.[12] In Boston, when traffic

[10] *The Providence Journal* (December 16, 1952).
[11] Kansas City Public Service Company, *Annual Report 1953*, p. 8.
[12] Data supplied by the American Transit Association.

decreased one-third from 1950 to 1960, operating costs and fixed charges increased a third.[13]

Increased traffic congestion has caused a greater expenditure of fuel and labor time without any corresponding increase in passenger revenues. It has been estimated that delays in downtown traffic absorb at least 18 per cent of total vehicle running time.[14] The transit company of Seattle reported that an increase in average speed of one mile per hour decreased operating costs at least 10 per cent. Traffic delays in Baltimore cost the transit company an estimated $100,000 per year, and in Milwaukee a 5 per cent slowdown in running time required 45 additional vehicles and additional annual expenses of $750,000. Additional running time in Philadelphia after the war resulted in extra wage costs of $1.2 million a year.[15]

Taxes have also been an added burden for the transit industry. In former years when transit enjoyed a monopoly of local transportation, the municipality required the payment of franchise taxes for the privilege of using the streets. In most cases these taxes are still levied despite the highly competitive conditions to which the industry is subject. Analysis of taxes paid by 100 of the major privately operated transit companies in 1963 revealed that the tax bill was 9.1 per cent of operating revenues. This average was exceeded in many cases, however, and 27 companies paid taxes of 12 to 20 per cent of total operating revenues. The highest tax figure in relation to operating revenues was 22 per cent.[16]

In 1963 the entire industry paid total taxes of $79 million, of which 56 per cent were federal and 44 per cent were state, county, and local taxes. (See Table 16.) It is estimated that about 25 per cent of all taxes paid by transit

[13] Mass Transportation Commission, Commonwealth of Massachusetts, *Boston Regional Survey—Public Transportation,* 1962, p. 21.
[14] The same, p. 62.
[15] Data supplied by the American Transit Association.
[16] The same.

to the combined tax collecting agencies are franchise type taxes imposed by local communities. These taxes are fre-

TABLE 16. Transit Taxes in 1963

Type of Tax	Amount	Per Cent
Federal taxes (total)	$44,014,000	55.8
Income taxes	11,120,000	14.1
Other federal taxes	32,894,000	41.7
State, county and local taxes	34,906,000	44.2
Total Taxes	$78,920,000	100.0

Source: American Transit Association, *Transit Fact Book, 1964,* p. 5.

quently based on gross revenues at a rate of from 2 to 5 per cent, regardless of the financial position of the company. In addition to the franchise tax, some cities tax their transit companies to support traffic police, public parks, street cleaning, snow removal, and bridge construction.

The burden of public payments is usually substantial. In one year the St. Louis transit company paid over one million dollars in gross receipts taxes alone, and under the existing tax structure, more than 50 per cent of the additional net revenue obtained from a fare raise is absorbed by increased taxes. Los Angeles transit lines paid over a half million dollars a year to the city for rights to operate over municipal streets and for maintaining the paved center strip between street car rails.

Rising total costs and declining patronage made it impossible for transit operations to make both ends meet with the five- and ten-cent fares in force over much of the country at the end of the Second World War. Transit fares in the United States averaged 7.1 cents in 1924 and were still only 7.2 cents in 1947. (See Table 17.) The difficulty in recent years of bringing fares into line with rising costs and longer hauls stems largely from a long history of underpricing public transportation.

In Boston, for example, it was stipulated in 1897 that the Boston Elevated Railway Company would maintain the fare at five cents for 25 years. The five-cent subway fare in New

TABLE 17. Trend in Average Transit Fares in the
United States

Year	Average Fare (cents)
1924	7.1
1925	7.1
1926	7.1
1927	7.0
1928	7.1
1929	7.2
1930	7.1
1931	6.8
1932	6.7
1933	6.6
1934	6.6
1935	6.5
1936	6.6
1937	6.6
1938	6.6
1939	6.7
1940	6.7
1941	6.7
1942	6.8
1943	6.9
1944	6.9
1945	6.9
1946	7.0
1947	7.2
1948	8.2
1949	9.3
1950	10.0
1951	11.0
1952	12.0
1953	13.1
1954	14.3
1955	14.8
1956	15.4
1957	15.8
1958	16.5
1959	17.1
1960	17.8
1961	18.2
1962	18.7
1963	19.0

Source: Data supplied by the American Transit Association.

York was long considered inviolate. In most cities it was at
least tacitly agreed that a ride on public transit should in-

volve no greater sum than a nickel, and public pressure supported by public utility commission rulings made it impossible for many years to reverse the long-established policy of low fares and low standards of service. Transit companies have generally had to show an actual operating loss before obtaining a fare increase, and action on requests for increases have been delayed in some cases for more than two years. A ten-month lag between application and effective date of fare increase is common.

The inflexible fare structure that had become crystallized in the transit industry over a long period of time finally gave way to the pressures exerted by postwar inflation, and between 1945 and 1954 the average fare for a transit ride in the United States doubled, rising from 6.9 cents to 14.3 cents. By 1963 the transit average fare rose to 19 cents. In 1964 there were 242 cities with a basic cash fare of 25 cents. Forty cities had a fare of 30 cents, and two charged 35 cents. Two cities had a basic cash fare of 28 cents. A total of 160 cities had fares of 20 cents. Despite the upward trend, however, 18 cities in the country still had a cash fare of ten cents. (See Table 18.) An increasing number of transit companies have zoned fares to reflect more adequately the growing length of rides and to attract the short-haul patronage that is discouraged by the high flat fares necessary to cover average costs. In the summer of 1954, over 160 transit companies in the United States and Canada were using zone fares in some form.[17]

Fares for school children are generally much lower than the basic rate, and they thus represent a substantial subsidy by the transit industry for education. In 1964 out of 502 companies reporting special school fares, 38 had eight-cent fares. Twenty-eight companies reported fares of seven and one-half cents or less.[18] As school fares do not cover transportation costs, the burden is shifted to other riders in the form of higher charges. In 1954 a total of 24 million Boston

[17] Norman Kennedy, "Would Zone Fares Help Transit Keep Its Riders?", *Traffic Quarterly* (April 1954), p. 142.
[18] Data supplied by the American Transit Association.

riders rode at the student rate of five cents, which was one fourth the regular fare of 20 cents. Student riders comprised 10 per cent of all passengers using the Metropolitan Transit System. Approximately half the transit deficit would have been covered if students had paid the full fare or if student transportation had been subsidized directly by the city.[19]

TABLE 18. Basic Cash Fares—Cities Over 25,000
Population, October 1964

Fare (in cents)	Number of Cities[a]
35	2
30	40
28	2
25	242
23	5
22	2
20	160
18	1
17	2
15	103
14	2
10	18

Source: Data supplied by the American Transit Association.

[a] There were 97 cities where no data were reported.

The problem of the transit industry, however, has not been merely one of adjusting fares upward to cover costs. The difficulty in raising fares is that every increase results in a loss of patronage. How much this loss amounts to is not clear, as fare increases have occurred at a time when transit patronage was falling due to other factors such as the elimination of driving restrictions at the end of the war, the resumption of automobile production, rising incomes, and the spreading of metropolitan area population.

In New York City the subway system experienced a 13.1 per cent decline in passengers during the year following the

[19] Information from the Metropolitan Transit Authority, Boston, Massachusetts.

increase in fare from five to ten cents in 1948. A survey made to determine what happened to the millions of riders lost to the transit system revealed that 7 per cent shifted to taxis, 40 per cent rode in private cars, and 53 per cent abandoned the ride altogether and either walked to their destination or stayed at home.[20] When New York transit fares were again raised from ten to fifteen cents in 1953, the New York Transit Authority incurred a further passenger decline of 11 per cent.

The Chicago Transit Authority conducted tests in 1953 to determine what effect a reduction in fare might have on the level of transit riding. On four Tuesdays in that year, when the cash fare was 20 cents and the token fare 17 cents for surface line transport and 18 cents for the elevated lines, a bargain fare of ten cents was introduced between the hours of 9:30 in the morning and 1:30 in the afternoon. The result was a 10 per cent increase in traffic over the system as a whole, but a decrease of 4.6 per cent in total revenue for that day. For the four Tuesdays of the experiment the estimated loss exceeded $50,000.[21]

In Minneapolis a similar test was made involving a 50 per cent reduction in fares for college and university students. This experiment brought no noticeable increase in travel. Other promotional fare experiments conducted in Detroit, Evanston, and Covington, Kentucky, led to the conclusion that whereas a fare increase would result in a decline in transit patronage, a decrease in the fare was unlikely to cause sufficient increase in the volume of travel to be economically feasible.

In coping with cost and fare problems, the transit industry has been subject to a system of fare and service regulation established in a period when the problem was to preserve fair competitive relations among carriers and to

[20] New York City Mayor's Committee on Management Survey, *Modern Management for the City of New York*, Vol. 1 (March 30, 1953), p. 149.

[21] Schroeder, *Metropolitan Transit Research*, Chap. 2 (December 20, 1954), pp. 47–48.

prevent exorbitant returns on transit investments. The regulatory set-up varies from state to state, and includes both public utility commission and local municipal regulatory bodies. Political rather than economic factors frequently govern decisions with respect to fares or the provision or abandonment of service. In some cases the political situation has led to long delays between the granting of wage increases and the approval of fare increases, with resultant financial difficulties for the carriers. Decisions that need to be made on the basis of technical questions of transportation policy must be decided by public officials often lacking technical competence.

The Financial Position of the Industry

As a result of increasing costs and declining revenues, a financial crisis has developed for transit in many cities, and the situation has made modernization impossible. Between 1954 and 1963, a total of 194 transit companies have abandoned operations, and many cities have been left without transit services. For many other companies, there appears to be no end to the vicious spiral of mounting costs and fares and declining service and patronage.

Transit revenues have been averaging close to $1.4 billion in recent years, but they have been more than offset by operating costs. Operating expenses in 1963 totaled $1,315 million, leaving net revenues of $75 million. After payment of taxes, there was a deficit of $4 million. Net income has been steadily decreasing since the late fifties and until 1962 net income had averaged less than 2 per cent of total revenues. (See Table 19.)

National figures, however, are in many respects misleading. Many transit companies still appear to be in a satisfactory financial position; others are incurring only minor losses. A few large companies account for a major share of the immediate financial problem of the industry. Of the 35 transit companies of major importance, 13 showed a

TABLE 19. Trends in Transit Operations, 1940–1963
(In millions of dollars)

Year	Operating Revenue	Operating Expense	Net Revenue	Taxes	Net Operating Income	Net Income as Per Cent of Revenue
1940	738.8	599.5	139.3	62.9	76.5	10.4
1941	802.4	645.9	156.5	67.0	89.5	11.2
1942	1,044.0	772.4	271.7	129.1	142.5	13.7
1943	1,299.5	936.9	362.6	187.1	175.4	13.5
1944	1,368.6	1,016.7	351.9	190.1	161.8	11.8
1945	1,386.2	1,071.7	341.5	165.2	149.3	10.8
1946	1,402.0	1,133.5	268.5	129.4	139.1	9.9
1947	1,395.8	1,243.1	152.7	105.3	47.4	3.4
1948	1,492.9	1,347.6	145.3	101.5	43.8	2.9
1949	1,495.4	1,341.9	153.5	89.4	64.1	4.3
1950	1,456.1	1,299.9	156.2	89.6	66.6	4.6
1951	1,476.6	1,334.6	142.0	95.8	46.2	3.1
1952	1,505.7	1,373.1	132.6	102.6	30.0	2.0
1953	1,517.2	1,374.0	143.2	97.9	45.3	3.0
1954	1,476.1	1,340.6	135.5	90.3	45.2	3.1
1955	1,430.7	1,280.8	149.9	94.0	55.9	3.9
1956	1,420.5	1,274.9	145.6	89.7	55.9	3.9
1957	1,390.1	1,265.1	125.0	88.0	37.0	2.7
1958	1,354.0	1,269.4	84.6	77.8	6.8	0.5
1959	1,381.1	1,269.9	111.2	85.4	25.8	1.9
1960	1,407.2	1,289.9	117.4	86.7	30.7	2.2
1961	1,389.7	1,295.8	93.9	77.2	16.7	1.2
1962	1,403.5	1,306.0	97.5	77.8	19.7	1.4
1963	1,390.1	1,315.2	74.9	78.9	−4.0[a]	—

Source: Data supplied by the American Transit Association.

a Deficit

TABLE 20. Net Income of Major Transit Companies
in the United States, 1963
(In thousands)

Company[a]	Population Served	Revenue Passengers	Net Income[b]
149	7,782	1,820,040	$20,880[c]
447	4,130	492,232	4,347
255	2,700	270,291	852
159	5,478	243,889	820
780	1,625	271,103	368[c]
50	1,304	n.a.[d]	18,173[c]
763	1,692	279,086	1,110[c]
699	2,026	112,501	8
384	1,750	n.a.[d]	2,816
202	1,167	91,750	599
675	763	141,082	7,212[c]
492	977	88,547	516
515	1,500	62,292	645
130	760	52,036	261
379	728	39,322	422
313 & 327-X	668	50,357	117[c]
743	667	n.a.[d]	n.a.[d]
18	n.a.[d]	36,656	11[c]
731	563	38,299	268[c]
601	632	86,767	n.a.[d]
669-A	855	41,185	1,534[c]
783	475	56,823	1,241[c]
547	700	29,377	120[c]
583	775	31,604	343
787	650	66,393	299[c]
26	1,780	26,573	106
118	n.a.[d]	42,158	440
385	n.a.[d]	26,509	201
752	n.a.[d]	5,212	306
162	469	26,746	276
684	800	20,268	38[c]
326	579	27,020	64
610	672	22,773	277
730	483	19,959	3
78	500	19,660	41

Source: Data supplied by the American Transit Association.

[a] American Transit Association code numbers have been used to maintain the confidential nature of the financial data for private companies.

[b] After federal income taxes.

[c] Deficit.

[d] Not available.

deficit in 1963, and the trend is toward increasing financial embarrassment.

The heaviest loss in 1963 was incurred by the New York City Transit Authority where the total loss from current operations was $20.9 million. This deficit was exclusive of debt service charges that are an obligation of the community. Another sizable loss was the $18.2 million deficit of Boston, exclusive of debt services. On the other hand, net income after federal taxes was in excess of $1 million for two companies, and 18 companies had net incomes ranging from $3,000 to $850,000.[22] (See Table 20.)

Transit operations are unprofitable wherever large investments in rail and rapid transit have been required to accommodate a high density of population and where as a consequence it has been difficult to adapt to changing urban patterns or to take advantage of motorized transportation. Deficits are also being incurred in old cities where the history of transit fares reveals a long period of low fares and high dividends that made it impossible to set aside sufficient revenues to renew equipment and to keep pace with technological change. Where population growth has been most rapid, it has been the practice to supply added service by bus, and transit companies have often been able to do this profitably by reason of the adaptability of the bus to the new urban growth and the ability of the bus to share right-of-way costs with other users.

The greatest deficit in the transit field for a long period has been in New York, which is also the city of highest population density. In Chicago, the second most densely populated city, public carriers had been in financial straits for two decades or more before the Chicago Transit Authority was established. The transit troubles of Philadelphia have also reached a precarious state, and the density of population in that city is fourth greatest in the country.

Of the thirteen cities in the United States with population densities in the central city exceeding 10,000 persons per

[22] Data supplied by the American Transit Association.

square mile, most are in financial difficulties or have recently been in such financial condition that it was necessary to resort to public ownership. These cities include New York, Chicago, Detroit, Boston, San Francisco, Pittsburgh, Milwaukee, and Philadelphia. On the other hand, cities with a fairly low density of population for their total size show relative prosperity in transit operations. These include San Antonio, Denver, Dallas, Kansas City, Cincinnati, and Houston. All of these cities have fewer than 8,000 persons per square mile and their transit operations appear to be in above average condition.

Rate of population growth has had an important impact on transit profitability. In cities where rapid expansion of population has produced substantial increases in transit patronage since 1940, there has also been a fairly good financial record. In cities that have experienced only moderate increases in population and a reduction of transit traffic since 1940, there have been substantial transit deficits. For example, traffic in New York, Chicago, and Boston in 1963 was 25 to 45 per cent below 1940. These declines were registered in areas of fairly stationary population. Population increases from 1940 to 1960 were only 4 per cent for New York City and 5 per cent for Chicago. In Boston, population decreased 10 per cent during the same period. But in rapidly expanding Houston and Dallas with population increases of 144 and 131 per cent, transit business was relatively prosperous.

The financial record of the transit industry has improved as fares have been adjusted upward. The New York City subway operating deficit was reduced some $30 million annually by the imposition of a 15-cent fare. The financial position of the industry as a whole or of individual companies, however, does not adequately reflect the nature of the transit problem or the outlook. For often a favorable showing conceals the fact that equipment is antiquated and service inadequate and unpleasant. Profitability or merely making both ends meet is too frequently accomplished at the expense of the rider.

Over an extended period of time, the attitude of a large

segment of the industry toward the rider has not been one to attract business, and in many cities today the degree of rush-hour crowding on public carriers and the acceptance of large numbers of standees as a condition for economic survival indicate the obstacles to future growth.

For many years obsolete and antiquated equipment has discouraged transit patronage. Of the 9,010 passenger cars operated by the transit systems in Boston, Chicago, Cleveland, Philadelphia, and New York in 1960, one third were built before 1931.[23] In Chicago at the beginning of 1945, the average age of street cars then in service was 31 years, and the average age of rapid transit cars was 39. In other words, the latter cars had been purchased, on the average, in 1906. As a result of the large-scale program of modernization by the Chicago Transit Authority, buses were substituted for electric cars of ancient vintage, bringing average age for all surface equipment down to six years in 1954 compared to nearly 29 in 1945.[24]

Approaches to the Transit Problem

A variety of efforts have been made to overcome the financial and operating difficulties of the transit industry. Among these have been measures designed to develop off-peak patronage, to speed up transit vehicles, and to reduce taxes and franchise requirements. In a number of cities resourceful transit companies have built up charter bus operations to utilize idle equipment in off-peak hours. Fleets of buses are rented to schools for educational trips. Buses are chartered for a variety of business purposes, and for sightseeing, conventions, and sports events. These and many other charter operations are helping to alleviate the off-peak problem.

One experiment eliminated fare collection in the down-

[23] Institute of Public Administration, *Urban Transportation and Public Policy,* 1961, New York, p. II–69 (mimeo.).

[24] Schroeder, *Metropolitan Transit Research,* Chap. 5 (May 19, 1955), p. 45.

town area to speed up service. Another approach has been to reserve exclusive lanes for buses during peak periods. Express services are provided in a large number of cities to speed rush-hour travel. Atlanta, Georgia, provides express service on 25 routes and patronage has increased steadily since the inception of the first expressway operation in 1955. The number of weekday passengers increased three and a half times between 1955 and 1964.[25]

In Boston, it has been concluded that the declining trend in transit riders is not inevitable, that frequency of service is more important than lower fares as a means of increasing patronage, and that selected improvements in frequency can be self-sustaining. For commuter railways, however, it was concluded that while better service will increase patronage, it will not eliminate deficits. The cost of providing better service was barely offset by revenue increases, and higher fares would not offset the deficits because the large increases called for (estimated at 50 per cent) would inevitably discourage users.[26]

A number of cities have taken steps to reduce taxes on transit. The gross receipts tax has been eliminated or lowered in many cities, and elsewhere transit companies have been relieved of paving between rails and removing snow along street car tracks. The Houston transit company has been granted a reduction in the gross receipts tax from 3 to 2 per cent, which means a saving of about $100,000 a year for the company. In Portland, Oregon, the 5 per cent gross receipts tax was reduced to 1 per cent, and the city waived the requirement that the company must pave over abandoned street car tracks. Grand Rapids, Michigan, permanently discontinued its 3 per cent gross receipts tax. The transit company in Washington, D.C., has been relieved of sanding icy streets, a franchise requirement that has been in effect in many cities despite the fact that the streets are shared by all types of motor vehicles.[27]

[25] From the American Transit Association.
[26] Mass Transportation in Massachusetts, Mass Transportation Commission of Massachusetts, July 1964, pp. 1–3.
[27] Data supplied by the American Transit Association.

Many cities and states have joined in a more extensive quest for financial and administrative solutions to critical mass transportation problems. The San Francisco Bay Area Rapid Transit Commission, created by the California legislature in 1951, recommended a master rapid transit plan and concluded that the transportation requirements of the area could not be handled by private vehicles alone. If the estimated passenger movements in 1970 were to be satisfied under the present pattern of regional transportation, an estimated total of 48 lanes of freeway would be required to handle anticipated peak-hour volumes at the principal gateways to the central metropolitan area. Ten to 12 lanes would be required to accommodate passenger cars crossing the Bay.[28]

The report of the Commission recommended a rapid transit system that would be a part of the unified network of local and interurban transportation. As a result, the Bay Area Rapid Transit District is building the most advanced rapid transit system in the world. The system will have about 75 miles of double track, including 16 miles of subways and tunnels, 31 miles of elevated structure, and 24 miles of surface routes. Completion date is 1971, at a cost of $1 billion. Construction of the transit system has been made possible by a $792 million bond issue voted by residents of San Francisco, Alameda, and Contra Costa counties in 1962; the remainder is to be financed through revenue bonds.

Specially designed lightweight electric trains will operate on the Bay Area's rapid transit system. They will be completely automatic, their operating controls governed entirely by a centrally located electronic computer to provide greater safety, operating efficiency, and economy. Trains will travel at top speeds of more than 70 miles an hour. Schedule speed, including station stops, will be ap-

[28] Report of the California Legislature Senate Interim Committee on San Francisco Bay Area Metropolitan Rapid Transit Problems, *Mass Rapid Transit in the San Francisco Bay Area* (March 1953), pp. 33–34.

proximately 50 miles an hour. Plans call for a computer-tabulating device to automatically record each passenger's entrance and exit through the station turnstiles, utilizing individually coded credit cards. The Bay Area traveler will be billed automatically for his total mileage at the end of each month. Newly developed currency change-making devices also will be installed at each station to facilitate collection of cash fares.[29]

A further step toward solution of urban transportation problems has been taken in the New York metropolitan region. In 1961, the governors of Connecticut, New Jersey, and New York met to discuss the region's transportation problems and subsequently announced the appointment of a Tri-State Transportation Committee.[30] The Committee is responsible for conducting a broad-scale examination as well as making recommendations for meeting the region's immediate and long-term transportation needs.

The tri-state metropolitan region, consisting of portions of the states of Connecticut, New Jersey, and New York, encompasses 8,000 square miles extending from New Haven, Connecticut, to Trenton, New Jersey, and is the nation's largest urban complex. The present population of 17 million is expected to increase to 23.5 million by 1985. In the face of this anticipated growth, the Committee is undertaking detailed analysis of region-wide mass transportation services, highway systems, freight handling, aviation facilities, and access to major commercial airports.

The Committee draws experts on loan from the highway departments of the three states, the Port of New York Authority, the New York–New Jersey Transportation Agency, and the Bureau of Public Roads, as well as from private sources. To facilitate the financing of the Tri-State

[29] "Rapid Transit for the Bay Area," Bay Area Rapid Transit District, San Francisco.

[30] By 1965 the three states passed legislation to transform the Committee into an interstate compact agency known as the Tri-State Transportation Commission.

program, the state of New York appropriates the necessary funds. New Jersey and Connecticut reimburse New York for their respective shares of the cost of operation.

Another comprehensive study of the transit problem was undertaken in Washington, D.C., under the direction of the National Capital Planning Commission. This study, supported by congressional appropriations, was based on a complete land-use survey of the Washington metropolitan area as a basis for determining future transit requirements.

In 1960 Congress created the National Capital Transportation Agency, which is responsible for a regional network of rail transit facilities focusing on the downtown area and extending into Virginia and Maryland. The initial development of the system involves 25 miles, 13 of them underground, with a total of 29 stations. Provision is made for co-ordinated bus services at suburban terminals to extend the use of the system over a wide area. In addition, suburban commuters would rely on parking garages near rail stations. It is anticipated that ultimately the network of rail routes will comprise 83 miles. The initial phase of the construction would require estimated outlays over $431 million; subsequent extensions might require an additional sum of $400 to $600 million.[31]

Organizational Arrangements

In many of the largest cities of the United States, transit service under private ownership has not been financially feasible, and various types of public authorities and boards have been established to take over bankrupt transit companies. These new arrangements have eliminated inflated values and have afforded some measure of tax relief and tax support. As of mid-1965, publicly owned transit systems accounted for approximately 55 per cent of the urban transit industry. Transit authorities are operating in

[31] Selma J. Mushkin and Robert Harris, "Transportation Outlays of States and Localities: Projections to 1970," The Council of State Governments, Chicago, 1965, p. 25.

most major cities, including New York, Chicago, Los Angeles, Detroit, Cleveland, San Francisco, and Boston.[32]

In St. Louis the Committee on Transit Ownership appointed by the Mayor concluded that public ownership of transit was the only solution to the transportation problem in that city. In the words of the committee, "if private industry is unable to furnish an adequate mass transportation system under conditions which exist and which are likely to continue to exist for a substantial period of time, the municipality not only has the right but is under the duty and obligation to provide such transportation."[33]

Among the reasons cited by the committee for recommending public ownership was the fact that under private ownership a substantial part of the earnings of the Public Service Company must be used to pay state and federal income taxes. Under public ownership these would not have to be paid but could be used to improve service and fortify the financial position of the system. An additional tax advantage would be the elimination of federal taxes on gasoline; and public ownership would mean no further need to earn a return on transit investment.

A further reason for suggesting public ownership was the fact that St. Louis has no control over the fares charged or service provided by the Public Service Company as this authority is vested in the State Public Service Commission. It was felt that this divorcing of regulatory authority from the jurisdiction of the city was an unnecessary obstacle to acceptable rate and service policies. A publicly owned authority, in the judgment of the committee, would be more responsive to rate and service matters because it would be closer to the problem than an agency concerned with state-wide public utility matters.[34]

[32] Other cities with a population of over 500,000 where transit lines are publicly owned include St. Louis, Dallas, Pittsburgh, San Antonio, Seattle, and Memphis.

[33] *Majority Report of the Mayor's Transit Ownership Committee*, St. Louis, Mo. (October 15, 1952), p. 14.

[34] A referendum on a county-city transit authority was voted on in 1955 and failed.

The decision to resort to a public authority in Chicago was reached after a number of fruitless efforts to reorganize the surface and rapid transit systems as a private company. "For more than fifty years various actions taken, both by private interests and by the City, in an effort to reorganize and procure an adequate transportation system, have all failed."[35] The solution was a partially self-supporting operation rather than an operation subsidized by tax levies.[36] The objective was to unify the transportation system in the Chicago metropolitan area by acquiring facilities or by agreeing with other local transit companies and railroads to set up joint fares and joint use of facilities. The modernization of existing facilities and the purchase of additional transportation properties were financed by the sale of revenue bonds to be repaid solely out of operating revenues.[37]

The Authority was given the power to fix fares at a level sufficient to pay all expenses. These include payments to the city for the use of certain city-owned properties and for covering costs of depreciation and modernization. The Authority does not pay real estate, federal income and excise taxes, or state gasoline taxes, and it benefits from the fact that both of the city-built subways are made available without payment of a specific rental. Transportation equipment for the subways was purchased by the city and the cost is being repaid by the Authority over a period of 31 years without interest. The Authority obtained its capital through the issuance of $105 million of revenue bonds, of which $87 million was used to purchase the properties.

[35] *People* v. *Chicago Transit Authority,* 392 Ill. 77, 88–89, in Werner W. Schroeder, *Metropolitan Transit Research,* Chap. 4 (March 31, 1955), p. 17.

[36] The Chicago Transit Authority, established in 1945, is operated by the Chicago Transit Board, which is responsible to the Illinois General Assembly. The board consists of seven members, three appointed by the Governor of Illinois and four by the Mayor of Chicago for a term of seven years.

[37] Chicago Transit Authority, "Local Transit in Chicago," *Traffic Engineering* (September 1952), pp. 457–73.

In the first seven and a half years of operation, some $98.5 million was spent for modernization of plant and equipment whereas only $46 million of capital funds had been spent in 38 years prior to Transit Authority administration.[38]

Except for the indirect financial aids provided by the city, as noted above, the Chicago Transit Authority earns revenues sufficient to meet all charges, including depreciation. It does this despite the large sums spent for modernization. Additional outlays have included purchase of the Chicago Motor Coach Company and the acquisition of railroad right-of-way formerly leased by the Authority. However, an extension of rapid transit service west from the downtown district authorized by the voters of Chicago has been financed through the sale of general obligation bonds. The new facility, operated by the Authority, is an extension of the Dearborn subway connecting with the rapid transit railroad in the center mall of the Eisenhower Expressway.[39]

The physical and financial accomplishments of the Chicago Transit Authority were made possible partly through effective management that consolidated three separate companies that had previously competed with overlapping services. Modernization of plant and equipment, together with the elimination of duplicating staffs and services, made it possible to reduce staff and effect substantial savings. The Chicago system is not completely self-supporting and must depend on general tax support for new investment in fixed plant. However, the new arrangement provided a successful approach to operating economy and the rehabilitation of rolling stock. The financial outlook, while not good, has been greatly improved.

Public transportation in Boston was furnished by the Metropolitan Transit Authority from 1947 to 1964. In the latter year the Massachusetts Bay Transportation Authority

[38] Schroeder, *Metropolitan Transit Research,* Chap. 5 (May 19, 1955), pp. 5, 6.
[39] Formerly the Congress Street Expressway.

was created to expand the metropolitan transit area to an additional 64 cities and towns. The new Authority was empowered to construct rapid transit lines, to make contracts with the railways for commuter service, and to participate in comprehensive planning activities of the Boston metropolitan region.[40]

Prior to establishment of the metropolitan authority, public ownership in Boston was already a quarter century old, and a long history of dubious financial arrangements seemed destined to lead to further difficulties. The city of Boston built the first subway in the United States in 1897 and chartered the Boston Elevated Railway to operate subway and elevated lines. When inflationary conditions after the First World War made the position of the company untenable, transit was placed under the control of a Board of Public Trustees. Deficits were made payable by the various communities served in proportion to the number of riders from each community. However, a gradual increase in fares from five to ten cents permitted the system to operate without a deficit until 1930.[41]

Despite the heavy traffic of the Second World War, the transit system of Boston encountered deficits in 1944, 1945, and 1946. When transit was unable to pay real estate taxes and interest, Boston decided to establish the Metropolitan Transit Authority. The new agency was directed to provide rapid transit, rail, bus, and trolley coach services for the 14 cities and towns in the metropolitan district. It was relieved of federal income and excise taxes and the guaranteed dividend to stockholders (amounting to over $600,000 per year) but was required to pay subway rentals, local real estate taxes, and interest on the bonded debt. Power to set rates was conferred on the Board of Public Trustees rather than on the Authority, and failure

[40] Mass Transportation Commission, Commonwealth of Massachusetts, "Mass Transportation in Massachusetts," Directed by Joseph F. Maloney. July 1964, pp. 116–17.

[41] Under the new set-up the owners were guaranteed dividends in the amount of 6 to 8 per cent.

to obtain consent for needed fare increases led to renewed deficits, which in the first six years amounted to more than $31 million.[42]

Legislation passed in 1954 transferred the fixed charges for debt retirement and subway rental from the transit system to the 14 cities and towns in the Boston area, and the Authority was left with responsibility for operating costs only. It is doubtful, however, that even these charges can be met by fare revenues at existing levels of traffic, and any future plant expansion would have to be financed from other sources.

A report on the transit problem of Boston has concluded that "public policy premised on the assumption that transit operations can be self-supporting is no longer realistic," as evidenced by the more than twenty years of annual deficits sustained by the transit system in Boston. "As long as this myth of transit solvency persists, deterioration of mass transportation services and the areas served by them will be inevitable."[43] It was further concluded that "the unworkable policy of attempting to finance MTA service from fare revenues should be abandoned and there should be a basic reallocation of transit costs among those benefiting from the existence of the system."[44] The report opposed further fare increases, and suggested instead that not only fixed charges but future operating losses as well should be met from sources other than the users. This was accomplished by assessing part of the operating losses on towns not then included in the transit district but which maintained connections with the Metropolitan Transit Authority system. The remaining operating losses were paid out of the general funds of the Commonwealth. Further financial support for transit under the Massachusetts Bay Transportation Authority includes a state-wide increase of 2 cents in the cigarette tax.

[42] Warren H. Deem, *The Problem of Boston's Metropolitan Transit Authority* (1953), p. 42.

[43] The same, pp. 53, 68.

[44] The same, p. 57.

The difficulties of the bankrupt Cleveland Railway Company were settled by municipal purchase of the system in 1942, and both the purchase price and the time of purchase were to prove salutary for the new public agency. At first the transit operations of the city were made part of the Department of Public Utilities, but buyers of the revenue bonds issued to finance acquisition of the system favored separate management under a board. In 1943 a charter amendment led to the establishment of a three-member Transit Board. Powers of the new board were circumscribed, however, by the stipulation that approval by the City Council was necessary for all capital expenditures or contracts involving more than $10,000.

In 1949 Cleveland made the decision to construct a rail rapid transit route and other improvements to the transit system. The Reconstruction Finance Corporation made a tentative commitment to purchase revenue mortgage bonds, provided that the city charter would be amended to provide for a Transit Board with complete authority over the system. The amendment, which removed the financial controls from the City Council, was approved.[45] The board has no taxing power and must fix fares at a level to keep the system on a self-sustaining basis. Beginning in 1958 such municipally owned transit systems were exempted from taxes as a result of a decision of the Ohio Supreme Court.[46]

Cleveland has found it possible to finance some extension of rapid transit with its own revenues. This was made feasible by the fact that 13 miles of electrified track on a depressed right-of-way, built many years ago as a Union Railroad Terminal project, was available at low cost. The financial success of Cleveland transit can also be traced

[45] The five paid members of the Transit Board, not more than three of whom may belong to the same political party, are appointed by the Mayor with the approval of the City Council for overlapping terms of five years. They may not hold any other public office.

[46] *Passenger Transport,* February 2, 1958.

to the low price paid by the city for the bankrupt properties in 1942, and to the heavy war-time traffic that helped write off the debt by 1952. The $35 million cost of 1.5 miles of subway in the Cleveland downtown area, however, had to be supported by a general obligation bond issue.[47] Today the Cleveland Transit System operates one of the fastest transit lines in the country, bringing people downtown in 15 minutes from points as far as seven miles away.

In New York, despite a volume of business totaling six million passengers per week, the financial condition of transit is the worst in the country.[48] The financial difficulties of the New York subways led in 1953 to the establishment of the New York City Transit Authority, which was given the power to "acquire, build, lease, maintain, finance and operate municipal transit facilities and to fix fares," thereby removing the transit deficit from the budget of the city where it had competed for public funds with other needed public undertakings. The Authority assumed control of the $1.7 billion city-owned subway, elevated, and bus lines operated by the Board of Transportation.[49]

A fare increase from 10 to 15 cents made it possible to cover operating costs during the first year of operation (despite an 11 per cent loss of traffic) and a loan from New York City provided funds for the Authority to proceed with a capital improvement program. This program involved a $670 million six-year plan for rehabilitation of existing facilities, extensions of the system, and new equipment and power plants. Most of the funds to pay for the work were to come from special borrowing authority for

[47] Norman Kennedy, *The Management of Publicly-Owned Transit Systems in Cleveland, Ohio, and in Toronto, Ontario, Canada,* Institute of Transportation and Traffic Engineering Technical Memorandum No. B-18, November 11, 1953, mimeo.

[48] Number of passengers from New York City Transit Authority, *Budget Data and Transit Facts,* 1964–65 issue, p. 5.

[49] Private lines continue to be operated under franchise granted by the Board of Estimate.

city transit improvements authorized by an amendment to the State Constitution.[50]

The Federal Role

The most significant new source of support for transit has been the recent entrance into this area by the federal government, first in the form of demonstration grants and loans to improve service and more recently through a capital grant program. In 1961 Congress promulgated the Housing Act of 1961, which specified that mass transportation planning should be an integral part of urban planning. The types of planning which may be assisted include the preparation of comprehensive urban transportation studies, means of alleviating traffic congestion, and methods of reducing transport needs.[51] At the same time, a congressional appropriation of $42.5 million was provided to carry out a program of loans and grants for improving mass transportation services.[52]

The program of demonstration grants in the 1961 Act has been characterized by a diversity of small projects. The first grant under this program was made to Detroit, Michigan, to help improve the city's bus service along a 14-mile bus route which connects downtown with the northwestern part of the city. The experiment cost one third of a million dollars. Two thirds of the cost came from the federal government and the remaining third from the city. The heart of the plan was to increase bus service up to 70 per cent during rush hours as well as in off-peak hours

[50] Voters in 1951 amended the state constitution to allow the city to borrow $500 million beyond its debt limit for the trunk line construction program. New York City Transit Authority, *First Annual Report 1953–1954*, pp. 24–25.

[51] Housing and Home Finance Agency, *Federal Laws: Assistance to Mass Transportation*, pp. 3–4, as quoted in George M. Smerk, *Urban Transportation: The Federal Role*, Indiana University Press, 1965, p. 149.

[52] Housing and Home Finance Agency, News Release OA–No. 62–174.

on Saturdays and Sundays to increase transit patronage. The result was a 12 per cent increase above the base period.[53]

The sum of $3.1 million in federal funds was granted to the Southeastern Pennsylvania Transportation Compact to operate low-fare, improved commuter service between the downtown district of Philadelphia and suburban Bucks and Montgomery counties. The project's remaining cost of $1.6 million was contributed by the participating counties. Service improvement at reduced fares commenced in 1962 on the Reading's Lansdale and Hatboro electrified suburban lines as well as on the Levittown section of the Pennsylvania Railroad. This service was supplemented by the provision of low-fare, improved bus transportation feeding the railroad suburban stations.

In 1963, Washington, D.C., received a grant from the HHFA to test the effectiveness of minibuses circulating within the central business district on a fixed route with frequent schedules. Buses seating 12 to 18 passengers are being provided and the fare is five cents. The total cost of the project was $240,000, of which two-thirds was contributed by the HHFA, and the remainder by the D. C. Transit System, which operates the buses.

The minibus system, composed of ten small buses, operates every two and a half minutes from 10 a.m. to 6 p.m., Mondays through Saturdays. Prior to Christmas, additional service was provided to meet heavy evening demand, and during the month of December 1965, the minibus system attracted a quarter million riders. Patronage has averaged 9,618 per day.[54] The federal grant was discontinued in November 1964 and since then the operations have been carried on by the D. C. Transit System.

Under the loan program of the Housing Act of 1961, the Chicago Transit Authority received a loan of $7.5 million from HHFA. The loan was to cover half the total cost of $15 million for the purchase of 180 new light-

53 Smerk, *Urban Transportation: The Federal Role,* p. 155.
54 *Passenger Transport,* January 22, 1965.

weight, electric rapid transit cars for the CTA's Lake Street and Congress-Douglas subway-elevated operations. The loan is to be repaid over a period of ten years and bears an interest rate of 3⅝ per cent.[55] It is estimated that 38.5 million revenue passengers will benefit from the service yearly.

With the various loans and grants for mass transportation projects made available under the Housing Act of 1961, federal participation was aimed mainly at experimentation. It was clear, however, that assistance on a regular basis for large-scale urban mass transportation projects was urgently needed. Recognition of this need came in the spring of 1962 when President Kennedy in his Transportation Message to the Congress called for federal financial assistance for local urban mass transit systems. This assistance became a reality with the passage of the Urban Mass Transportation Act of 1964. The new law authorized $375 million to be used to finance two thirds of the total cost of transit projects over a three-year period.

The Act provided funds for capital investment, such as purchases of land, right-of-way, parking facilities, rail cars, stations, and terminals. In order to be eligible, an area seeking aid must work out a co-ordinated urban transportation system as part of a comprehensive plan for the region. Since the Federal-Aid Highway Act of 1962 also requires comprehensive planning, a step forward was made toward co-ordination of federally aided highway and mass transportation projects.

Other major provisions of the Act include the continuation of research, development, and demonstration projects as well as a program of loans for equipment previously provided for under the Housing Act of 1961. In addition, the HHFA Administrator may initiate demonstration or research projects independently as well as by contract with public or private sponsors.[56]

---

[55] Smerk, *Urban Transportation: The Federal Role,* p. 170.
[56] Public Law 88–365, 88th Cong. S.6, July 9, 1964. Urban Mass Transportation Act of 1964.

The Problem of Railroad Commuter Service

The problems associated with the movement of commuters by railroad are similar to those that have been reviewed in connection with municipal rail and bus operations. Railroad commutation is important in only a few large eastern and midwestern cities, but in those locations it plays a substantial part in the total mass transportation system. Commuter passengers comprise more than half of all rail passenger business and are responsible for a sizable share of the financial difficulties of the railroads. The passenger service deficit on the nation's railroads in 1963 was $398 million, and 33 per cent of net railway operating income from freight was absorbed in making up this deficit.[57] (See Table 21.)

Rail commutation has fallen off rapidly despite the growth of urban population. (See Chart 14.) Trends in the location of business in the city have meant that downtown terminals once centrally located have become inconvenient in relation to commuter destinations. And in the suburbs the sprawl of residential development has meant that large areas are not served by rail lines and are much better adapted to the flexible services of bus or automobile.

Two additional problems have played an important part in the troubles that have beset the railroads. First the predominance of home-to-work travel means that peak-load traffic is especially heavy in relation to travel in off-peak hours. Second, fares have been maintained at unremunerative levels over a long period. Revenue per passenger mile on commuter lines rose from 1.1 cents in 1922 to 1.12 cents in 1947. An increase has occurred since that time, but in 1955 the fare was still only 2.12 cents and in 1963, 3.17 cents per passenger mile. In the Chicago area some

---

[57] Deficit figures are based on fully allocated costs as determined by Interstate Commerce Commission formula. They exaggerate the passenger deficit situation, and in some cases commuter service is covering direct costs.

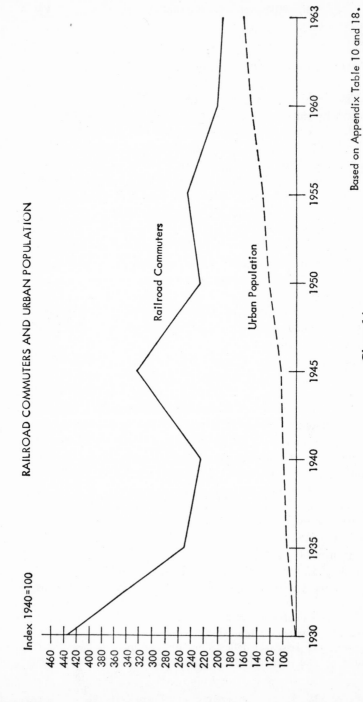

RAILROAD COMMUTERS AND URBAN POPULATION

Index 1940=100

460
440
420
400
380
360
340
320
300
280
260
240
220
200
180
160
140
120
100

1930   1935   1940   1945   1950   1955   1960   1963

Railroad Commuters

Urban Population

Based on Appendix Table 10 and 18.

Chart 14

TABLE 21. Net Railway Operating Income[a]
(In millions)

Period	Freight Service	Passenger Service	Total[b]
Average:			
1936–41	$ 891.4	$244.9[c]	$ 649.9
1942–45	991.6	208.3	1,200.7
Year:			
1946	759.7	139.7[c]	620.1
1947	1,206.4	426.5[c]	780.7
1948	1,561.0	559.8[c]	1,002.0
1949	1,335.5	649.6[c]	686.5
1950	1,547.7	508.5[c]	1,039.7
1951	1,622.9	680.8[c]	942.5
1952	1,720.1	642.4[c]	1,078.2
1953	1,812.8	704.5[c]	1,109.4
1954	1,543.1	669.5[c]	874.0
1955	1,764.3	636.7[c]	1,128.0
1956	1,764.7	696.9[c]	1,068.2
1957	1,645.9	723.7[c]	922.2
1958	1,372.8	610.4[c]	762.3
1959	1,291.9	543.8[c]	747.7
1960	1,069.0	485.2[c]	584.0
1961	944.8	408.2[c]	537.8
1962	1,119.2	394.0[c]	725.7
1963	1,203.6	398.4[c]	806.4

Source: 1936–53 data from Interstate Commerce Commission, *Transport Economics*, May 1956, p. 1; 1954–63 data from Interstate Commerce Commission, *Transport Economics*, May 1964, p. 5.

[a] Class I Railroads only.
[b] Includes relatively small amounts not related to freight or passenger service.
[c] Deficit.

of the monthly commutation fares to outlying points were as low as one cent per mile in the early fifties. Yet at some points traffic thinned out so much that it cost the railroad over five cents per passenger mile to provide the service for which it charged a penny a mile.[58] (See Appendix Table 18.)

[58] Stanley Berge, "Chicago Suburban Passenger Transportation," An Address before the Chicago Regional Planning Association, Chicago, April 22, 1952.

On the Chicago, Aurora, and Elgin Railway, serving the west suburban district of Chicago, more than 70 per cent of the passengers rode during the peak periods. Eastbound rush-hour traffic in the morning and westbound traffic in the evening was on the order of twenty times the traffic in the reverse direction. Estimates of revenues and expenses indicated that no net revenues could be earned at existing levels of fares even with the additional traffic that might be generated if modernization led to improved service.[59]

The Long Island Railroad, serving one of the most densely populated suburban areas in the country, has been in financial difficulties for a long period. The Long Island carries about one fifth of all the commuter railroad passengers in the country. Because it is primarily a commuter railroad, it has been unable to offset passenger service losses with freight revenues, and the resulting deterioration in equipment and service has been a twice-daily cause of unhappiness for New York commuters. On one Long Island run, elapsed time had increased from one hour and one minute in 1889 to one hour, thirty-three minutes for the same distance a half century later.[60]

About a decade ago, the problem of the railroad was described to the Interstate Commerce Commission as follows:

The stark condition from which the Long Island cannot escape is that it is faced by the ever-expanding competition of motor vehicles—private cars, buses, and trucks —and the New York City subways. Increases in passenger fares and freight rates cannot be relied upon to produce sufficiently large increases in revenues to alter materially the present level of earnings.[61]

[59] Charles E. DeLeuw, "Economic Aspects of Planning Rapid Transit in Metropolitan Areas," Paper presented at the Fifth Annual California Street and Highway Conference, Berkeley, California, February 6, 1953.

[60] New York Times (March 21, 1954).

[61] Memorandum to the Interstate Commerce Commission by the Long Island Transit Authority (October 2, 1952).

Rehabilitation of the railroad in accordance with recommendations of the Long Island Transit Authority has been undertaken through a twelve-year development program.[62] Purchase of new air-conditioned equipment and the modernization of other facilities have been accomplished by a combination of liberal borrowing, tax concessions on the part of the state, city, and county governments, higher fares, and the waiver by the parent Pennsylvania Railroad of interest, dividend, and principal payments on the present investment.

The railroads serving the New York metropolitan area operate their passenger services at sizable annual deficits ranging from $3 to $10 million, but passenger losses except in the case of the Long Island are made up by freight revenues. All commuter railroads are carrying fewer passengers today than they did in the past, but the impact of the motor vehicle has been most severe on the railroads serving New Jersey, with the exception of the Pennsylvania. During the period since 1950, the number of commuters using New Jersey railroads has declined 50 per cent, while the passenger loss on the Westchester lines has been approximately 25 per cent and there has been a slight decline on the Long Island.[63] The absence of direct rail links to mid-Manhattan and competition of the Hudson crossings of the Port Authority have cost the railroads on the west shore much of their business, as only the Pennsylvania offers high-speed service direct to mid-Manhattan.

A serious problem is the large volume of "urban" traffic that increases operating costs out of line with any possible return, requiring extra cars and crews for the full length of the trip despite a poor over-all occupied seat

[62] *A Plan for the Rehabilitation of the L.I.R.R.*, A Report to Governor Dewey by the Long Island Transit Authority (May 26, 1954).

[63] Data from tabulations of Planning Division, Port of New York Authority, March 1955; *Metropolitan Transportation—1980*, The Port of New York Authority, 1963, pp. 292, 339; and *Suburbs to Grand Central*, Institute of Public Administration, 1963, pp. 1–16.

ratio. Seat-mile costs are incurred for the full trip, but seat use is for a short distance, and the revenue return is low. While the number of long distance commuters continues to increase, the large number of remaining "urban" passengers reduces average distance per commuter to 14 miles. The Pennsylvania Railroad estimates that it costs twice as much per passenger mile to serve close-in customers as it does commuters, as the empty seats must be hauled the full route distance.[64] The New York Central and New Haven railroads have been incurring heavy deficits and for some years their services have been deteriorating despite the efforts made by state governments to aid them. It is claimed that the two railroads together could save more than $10 million a year simply by abandoning the commuter services.[65]

The downward trends in mass transportation, whether by local transit or commuter rail lines, are in part the product of revolutionary changes in economics and technology and shifting patterns of urbanization. Competition of the automobile and the spreading area of low-density development made possible by highway transportation have made it difficult for public carriers to maintain satisfactory standards of service. But combined with these basic factors have been the past errors of management and the persistence to date of outmoded public policies that have saddled mass transportation with an inheritance of obsolete facilities, antiquated regulations, and inequitable conditions of taxation and finance. One of the principal questions clouding the future of urban transportation is the role that can be expected of public carriers. The answer will have a significant impact on future patterns of urban development in the United States.

[64] Presentation by the Pennsylvania and Reading railroads to the Urban Traffic and Transportation Board, Philadelphia, May 21, 1954.

[65] Press release issued on January 6, 1964, by the Institute of Public Administration, New York City.

CHAPTER IV

Role of Public and Private Transportation

Current efforts to supply satisfactory standards of urban transportation service, as noted in the previous chapter, have approached each method separately, with the result that limited attention has been given to the question of the relative roles to be played by public transportation and private automobiles. Great efforts have been made and are being planned to accommodate the use of private passenger cars in major metropolitan cities. The difficulties of doing so, however, and the cost involved, present formidable obstacles, and existing public policies show limited promise of overcoming them. Public transportation, on the other hand, which in many cases seems the more logical way to serve cities of high population density, continues to lose patronage and receives relatively little attention in programs of modernization and development.

Public policy appears to be committed to placing major emphasis on the automobile to the neglect of public carriers. The question has been raised whether such a policy can succeed in providing the transportation services that cities require, or whether the result will merely compound the problems of congestion and fail to furnish adequate transport standards. The suggestion is frequently made that public policy should focus instead on the development of mass transportation, with the objective of reducing the number of vehicles using the city streets, and that such a course can provide the only long-run solution to the urban transportation problem. The basic question, then, is the relative roles that should be assumed by the automobile and by public carrier transportation to meet the demands for urban mobility in the future. The answers will differ as the nature and intensity of the problem differ, but many of the same factors will underlie the decisions that must ultimately be made.

Automobiles vs. Transit

If the growth in motor vehicle ownership and use indicated in Chapter II should actually take place, and if unlimited use of private vehicles were to be permitted, existing big cities would have to undergo a metamorphosis that would mean far greater dispersal and far lower densities of population than is the case today. This would be necessary to make way for the highways and parking facilities that would be required. The only other alternatives are that car ownership will not increase as projected, or that the provision of facilities for motor traffic will not be on a scale adequate to meet all needs at all times. In the latter case mass transportation facilities would have to be relied upon to accommodate a substantial proportion of peak-load home-to-work traffic.

The possibility that we shall redesign all of the high-density areas of large cities to the extent necessary to permit unlimited private transportation during peak hours seems in the near future to be both improbable and impractical. The trend will certainly be in that direction, but experience indicates that the change will be gradual. Otherwise we would not only incur a heavy loss of investment in urban areas, but we would face at once the problem of providing substitute urban plant and utilities elsewhere. The task of accommodating a rapidly expanding urban population will demand whatever can be salvaged from today's cities.

The second alternative, that projected increases in automobile ownership may not materialize, also seems unrealistic in a nation of automobile owners that can look with confidence to a period of continuing economic growth. Thus the third alternative, that the accommodation of private transportation will be on a lesser scale than necessary to permit unlimited peak-hour mobility, seems to be suggested. This would circumscribe the extent to which people could drive themselves to the same places at the same time, at least in the near future. Urban development would continue to give way to the demands of private transportation, but some densely developed areas considered worth preserving would stand their ground against the automobile. The outcome would be in the nature of a compromise between private and public transportation that would vary among cities in accordance with physical necessity, economic feasibility, and consumer choice.

A certain minimum of public transportation service is essential in the big city. Factors determining what this level of service may be include the number of persons not owning automobiles or for whom an automobile is not available, and persons whose age or physical condition does not permit driving. There is also a certain level of essential automobile use, since some areas are not served by transit, some people need their cars in connection with their work, and some cannot be persuaded to use public carriers for a variety of personal reasons. Many other urban residents who do not fall in any of these categories will use one form of transportation or another depending on considerations of economy, standards of transit service, traffic and parking conditions, and a variety of circumstances governing specific trips.

Granting that some people will have to use public transportation and some will have to use their automobiles, the question in large cities today is the extent to which it may be possible or desirable to induce more people in the in-between category to use public carriers; and to what extent improvements in public carrier service would increase the use of transit. Thus far, too little is known about the magnitude of this middle category of riders, the motivations of the urban traveler, and the considerations that might lead him to travel by one method or another. Nor is it clear what the relative costs to the city of accommodating individual and mass transportation would be, all costs considered. If a sizable number of people could be "won back" to mass transportation, the questions arise whether the shift would alleviate transportation difficulties and whether an increase in the number of transit riders at the peak would relieve or compound the operating problems and financial position of transit.

Increasing interest is being shown in the possibilities of expanding public transportation patronage, partly as a means of improving the position of the transit lines and partly in the hope of reducing the seemingly insatiable demand for new highways and the mounting congestion downtown. From the standpoint of the city, it is clear that the amount of street space required to furnish mass transportation is much less than that required to move comparable traffic volumes by private automobile. Because this is so, the conclusion has been drawn that a reduction in the number of automobiles operating on urban streets "is the practical, realistic, and only course to effective relief from street congestion."[1]

An extreme illustration is provided by the situation in New York City, where on a typical weekday 3.5 million persons enter Manhattan south of 52nd Street. Most of them arrive by railroad, subway, and bus. If everybody entered instead by private automobile, parking would require five levels of storage area occupying all of the usable building space in Manhattan from 52nd Street to the Battery. If only half came by auto, ten square miles of parking space would be required, plus a doubling of street, bridge, and tunnel capacity.[2]

If the objective is to maintain the highest possible densities in central cities, the advantage of mass transport over the private automobile is hardly disputable. It has been pointed out in support of mass transportation that 100,000 people per hour can be moved in one direction on two subway tracks whereas 20 four-lane highways would be

[1] John Bauer and Peter Costello, *Transit Modernization and Street Traffic Control* (1950), p. 18.

[2] C. McKim Norton, "Metropolitan Transportation," *An Approach to Urban Planning,* Gerald Breese and Dorothy E. Whitman, eds. (1953), p. 87.

required to carry the same load in private automobiles. But from an economic standpoint, the problem is much more complex than comparing the space requirements or load capacities for different methods of movement. The question of relative costs concerns not only the cost of transportation but the cost of urban living under various densities of development. The New York subway, despite its deficits, may be the cheapest way to move large numbers of New Yorkers. But the intensity of urban development that has been made possible by the subway system has involved costs that are a heavy burden on the city and its people.

A decision as to the proportion of urban traffic to be accommodated by automobiles and public carriers, then, needs to be based on more than the relative number of passengers that can be carried past a given point in a given time. There are many considerations of cost and service involved. Moreover, local conditions will often affect the decision. In New York, for example, it is a physical impossibility to accommodate the private automobile to any large extent, whereas in other cities it may be equally impractical to suggest a subway. In most cases, however, considerable latitude is offered in choosing between transit improvements and the provision of highway and parking capacity for automobiles.

Comparative Costs of Public and Private Transportation

It is frequently asserted that mass transportation is cheaper for the consumer than transportation by automobile. "The automobile is a very convenient form of transportation but an expensive one. A trip from home to work costs between 5 and 10 times as much as a trip on the transit system."[3] Elsewhere it has been stated that a Philadelphian who travels twelve miles round trip between

[3] U. S. National Capital Park and Planning Commission, *Moving People and Goods,* Monograph 5 (June 1950), p. 6.

home and work can save $327.50 per year by using public transportation instead of an automobile.[4] These estimates, based on total costs and average costs, present an oversimplified picture of the actual competitive cost relationships between automobile and transit.

From the standpoint of the consumer, cost competition between methods of moving is much more complex.[5] The comparative costs of driving differ over a wide range depending on a variety of circumstances. Taking all items into consideration, the average cost of operating a new car is approximately 12 cents per mile. This figure includes direct operating expenses—gasoline and oil, maintenance and tires —and fixed charges—depreciation, insurance, and license fee. (See Table 22, p. 125.) But the cost of automobile operation rarely conforms to this average for new cars. Different geographical locations mean variations in taxes and fees. Insurance cost varies with the type and value of the policies, or insurance may not be carried at all. There is also a wide range in car prices, hence in the annual cost of vehicle depreciation. The fact that more than half of all purchases are used cars rather than new ones means still wider variation in sales prices and annual depreciation charges. Other factors influencing cost include speed, road design and surface type, and the degree of traffic congestion.

These variations are significant, but of much greater influence on the competitive relations between transit and automobile is the fact that the consumer in making a choice rarely takes all these costs into consideration. The family car is used for a variety of purposes, including the trip to work, to the store, and for social and recreational purposes.

[4] G. M. Woods, *Public Transportation and Its Effect on Traffic Congestion* (June 1951), p. 2.

[5] The competitive situation discussed here is limited to consumer costs as they may influence the choice of transport method and does not involve total economic cost, including both consumer outlays and public expenditure. Relative economy in the latter sense is considered later.

TABLE 22. Annual Cost of Operating a New Car, 1965[a]

Cost Items	Cost in Cents Per Mile for 10,000 Miles
Variable Costs:	
Gas and oil	2.58
Maintenance	.68
Tires	.44
Total	3.70
Fixed Costs:	
Fire and theft insurance	.31
Property damage and liability	1.26
License fees	.24
Depreciation	6.26
Total	8.07
Total Cost	11.77

[a] American Automobile Association, from data supplied by Runzheimer and Company, 1965. These are the national average cost figures for a 1965 Chevrolet, eight-cylinder, Bel Air four-door sedan.

Once a car has been purchased only the out-of-pocket cost, the extra or marginal cost that can be assigned specifically to the trip in question, is significant in deciding whether the car should be used or left at home. The marginal cost of which the motorist is actually aware might include only the gas and oil needed to keep the automobile going, and occasional repairs. Today total out-of-pocket costs may amount to 3.7 cents per mile.

The competitive advantage the automobile enjoys because the consumer tends to consider only marginal costs is great. An even more important advantage is the ability of the car to accommodate extra passengers at no extra cost. When comparisons between automobile operating costs and the cost to go by bus or street car are made on a per-passenger basis, the cost per passenger may be significantly lower for the automobile. An automobile carrying two people might operate at an out-of-pocket cost of 1.85 cents per passenger per mile, and for four or more people the cost per passenger-mile could be less than one cent.

Use of an automobile, however, includes not only the cost of moving but the cost of parking, and the parking fee is often higher than all other costs combined. For a commuter who drives a ten-mile round trip per day the out-of-pocket cost of car operation is 37 cents, but the addition of a $1.00 parking charge raises total trip cost to $1.37. With two riders, however, the cost is reduced to 68.5 cents per passenger.

Comparisons of automobile costs with the cost of moving by public carrier under different circumstances introduce a variety of competitive situations. Varying costs of automobile operation must be compared with transit fares that also differ widely. There are flat fares, zoned fares, reduced rates for tokens or round trips, special rates for school children, and free rides for smaller children. Thus comparative cost advantages must be determined on the basis of specific trips and circumstances. A few hypothetical situations, however, will serve to illustrate the cost differentials.

A bus trip of one mile and return in a city with a flat fare of 20 cents would cost 40 cents for the round trip. Two members of a family could make the trip for 80 cents. If two people made the same trip by automobile, the out-of-pocket cost assignable to that trip would be approximately 7.4 cents, or 3.7 cents per passenger. If parking were provided at the curb for 10 cents, the round trip cost would be 17.4 cents, or 8.7 cents per passenger. A charge of 50 cents for parking would raise the cost to 57.4 cents—17.4 cents above the transit fare for one passenger but 22.6 cents below the cost for two people by bus.

For a trip that is short in distance and duration, then, the advantage is likely to be with the automobile, and the margin of advantage is increased as the number of passengers in the automobile increases. But parking costs may prove to be the deciding factor. A round trip of 12 miles with a bus fare of 25 cents each way would mean a total fare of 50 cents compared with direct operating costs of about 45 cents to make the same trip by automobile. But if parking charges added another 50 cents to the cost of using a car, the total would be approximately 95 cents for the round trip by auto compared with 50 cents by bus. If two members of the family were making the trip the cost by auto would be reduced to 47 cents on a per-passenger basis, 3 cents less than the bus fare per passenger. A parking

ROUND-TRIP COST OF AUTOMOBILE AND BUS RIDE
(TWO PASSENGERS)

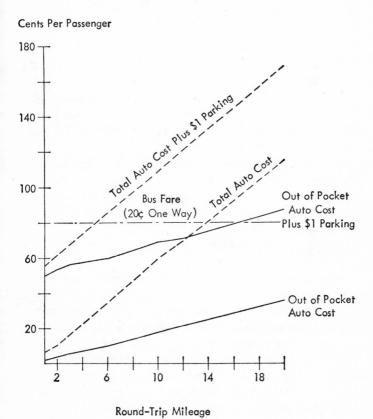

Based on Appendix Table 19.

Chart 15

charge of $1.00 would mean direct costs of $1.45 for one person to drive, or 72.5 cents per person for two.

The chart on page 127 provides an illustration of costs of travel per person by automobile and bus under a variety of assumptions with respect to trip length, fares, number of passengers, and cost of parking. Computations include total as well as marginal costs. Cost data indicate, for example, that a commuter making a round trip of 20 miles per day and paying $1.00 to park will have a transportation bill of $3.36 per day, assuming he is driving a new car costing 12 cents per mile to operate. If only out-of-pocket costs are considered, the round trip including parking would be $1.74.

If parking is provided free, either on the street or by the employer, total cost would be $2.36 and marginal cost 74 cents. These figures compare with bus fares that in the illustration vary from 30 cents to 80 cents per round trip. If two people were making the trip, driving cost per person might range from 37 cents to $1.68, while cost by bus would be 30 to 80 cents per person. Under these assumptions a wide range of cost differentials is revealed. It might cost twice as much per person to go by bus as to drive, or 11 times more to go by automobile than by bus.

Quality of Transportation Service

Competition between automobile and public transportation is more likely to be decided on the basis of service, however, rather than cost, and the fact that the automobile generally provides a superior quality of transportation without necessarily costing more, places it in a highly favorable position. The automobile provides flexibility of routing and schedules, savings in time, and complete service from any point of origin to any destination. Public transportation, on the other hand, is confined to specific routes and schedules, furnishing less convenience, and generally taking more time and effort per trip. The discomforts of transit riding that result from waiting, standing, changing, poor ventilation, and overcrowding add to the competitive advantage of the automobile.

Speed of service has been found to be of particular concern to the consumer, and this has placed transit at a considerable disadvantage. In the central business district, where the speed of the automobile is sharply curtailed by poor highways and heavy traffic, average driving speeds are nevertheless substantially greater than the transit vehicle that operates on the surface. In Cleveland it has been found that automobiles average 28 miles per hour downtown, whereas the transit vehicle that must stop to load and unload as well as cope with congestion makes no more than 11 miles per hour. Comparable figures for Washington, D.C., are 18 and 9 miles per hour. (See Table 23, below.) In Pittsburgh, the automobile is two and a half times faster than the bus, not counting the wait and the walk.

Rail rapid transit, freed from surface congestion but still slowed down by passenger stops, exceeds automobile speeds in New York with an average of 22 miles per hour compared to 21 for automobile. Express buses on expressways nearly approach automobile speeds on average city streets.

TABLE 23. Speed of Transit vs. Auto in the Central
Business District
(Miles per hour)

City	Auto	Transit
New York	21	22 (rail)
Chicago	20	16 (rail)
Philadelphia	18	15 (rail)
Los Angeles	21	17
Detroit	20	21
Baltimore	18	11
Cleveland	28	11
St. Louis	16	14
Washington, D.C.	18	9
Boston	16	12
San Francisco	22	32 (rail)
Pittsburgh	20	8
Milwaukee	22	14

Source: Editors of Fortune, *The Exploding Metropolis*, Doubleday & Company, Inc., 1958, pp. 60–63.

What people think about transit service is indicated in a study conducted by the College of Business Administration of Marquette University.[6] Of a thousand households interviewed in Milwaukee, 54 per cent said they never used mass transportation. Of those who used it, 30 per cent complained the ride was too slow, 15 per cent said the buses were too crowded, 18 per cent said they had to wait too long, and 13 per cent said it cost too much.

Altogether, 83 per cent of those interviewed said they thought transit service could be improved. Although 53 per cent of the families suggested more frequent service as a method of improving transit, only 34 per cent stated that they would increase their use of transit service if this improvement were made. On the other hand, 38 per cent said they would use transit more often if fares were lowered, and 27 per cent said they would use the service more frequently if faster service were provided. But 56 per cent of the respondents said they would still prefer to use their automobiles to go to work no matter what was done to improve transit.

Further light on the attitude of the public toward transit is supplied by a survey of automobile use in 1951.[7] When asked what they would do if the cars they used were not available, 37 per cent of the respondents said they would use another car, and nearly 11 per cent said they would walk. Only one third said they would use mass transportation.

Altogether 80 per cent of the people questioned specified some alternative method of transportation. These respondents were then asked what they would have done if their choice of first alternative had not been available. An additional 14 per cent said they would have walked. Twenty-eight per cent would have stayed at home or did not know

[6] *The City of Milwaukee's Mass Transportation Problem,* A Report by the Municipal Transportation Study Committee, with Survey by Marquette University, College of Business Administration (1955).

[7] Alfred Politz Research, Inc., "The Automobile in the Daily Life of the American Population," April 1951, unpublished.

DISTANCE AND METHOD OF TRIP TO WORK

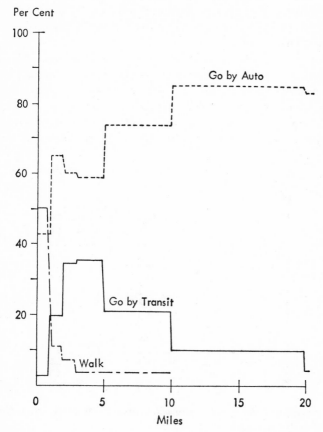

Based on Appendix Table 20.

Chart 16

what they would have done, 22 per cent would have used another car, 18 per cent would have used transit, and 10 per cent would have gone by taxi.

Growth of the urban area and the longer haul to work have favored the automobile as the factors of time saving and flexibility become more important with distance. At the other end of the scale, a large number of city trips are so short that they discourage transit patronage either because the time consumed in walking to the bus stop is a disproportionate part of total trip time or because the fare per mile for a short trip is large and the chances of a seat are small. In a study of trip lengths involved in journeys to work in incorporated places in six states, it was found that 31 per cent of all workers were traveling less than one mile to work and half were traveling less than two miles. Of the long-distance commuters, 6 per cent were traveling 10 to 20 miles. Among those who traveled less than one mile, 43 per cent went by automobile, and only 3.6 per cent by transit. Half of the group walked. The journey was appar-

TABLE 24. Passenger Trip Lengths by Distance and Method, Detroit, 1955[a]

Distance	Auto (Per Cent)	Mass Transit (Per Cent)
Less than 1 mile	22.7	7.5
1–1.9 miles	20.1	20.9
2–2.9 miles	12.8	16.1
3–3.9 miles	9.2	12.6
4–4.9 miles	7.8	10.6
5–5.9 miles	5.9	8.0
6–6.9 miles	5.3	7.5
7–7.9 miles	3.9	5.0
8–8.9 miles	3.2	3.9
9–9.9 miles	2.4	2.5
10 miles and over	6.7	5.3
Total	100.0	100.0

[a] *Report on the Detroit Metropolitan Area Traffic Study*, Pt. 1 (July 1955), p. 91.

ently too short to make transit attractive or feasible. Of those traveling the longer distances, 86 per cent drove, and fewer than 11 per cent used transit.[8] (See Chart 16, p. 131.)

In Detroit the use of mass transportation and automobile indicated a similar pattern. Of the automobile trips in the metropolitan area, 23 per cent were less than one mile. Only

[8] Thurley A. Bostick, Roy T. Messer, Clarence A. Steele, "Motor-Vehicle-Use Studies in Six States," *Public Roads* (December 1954), p. 111.

7.5 per cent of transit trips were in this category. Altogether 56 per cent of automobile trips were under three miles. (See Table 24.)

Rail Transport vs. *Bus*

In previous discussion of transportation trends, it has been seen that the volume of traffic moving by rail has declined over an extended period of time. In 476 cities of over 25,000 population, 429 have no rail transit service of any kind.[9] There has also been a long-run decline in movement of passengers by commuter railroad.

As urban street congestion has increased, the question frequently asked is whether this shift from rail to road has been predicated on sound economics or on the failure of public policy and the industry to achieve service adequate to attract and hold the traffic. Is a continuation of these trends inevitable, or is a revival of rail rapid transit possible, either in the form of subways or elevated railroads or extensions of railroad commutation lines?

Some cities are attempting to revitalize rapid transit. Cleveland, for example, has constructed new subway facilities, a rail transit system is under way in San Francisco, and improved rail commuter facilities are being sought in the New York–New Jersey and Philadelphia–South Jersey areas. In addition, the Toronto Transit Commission reports that the new Toronto subway "has demonstrated its ability to restore the downtown area to its rightful position as the heart of the city," and that "it has slowed down the flight to the suburbs."[10] Montreal and Chicago have also provided additional rail facilities to help solve their transit problems. But some view rapid transit and subway developments as "foredoomed to failure." It has been charged that Cleveland, "an otherwise solvent, profit-producing operation with a book value in excess of 50 millions of dollars,

[9] Data supplied by the American Transit Association.
[10] *Traffic Engineering* (December 1954), p. 114.

mortgaged itself to the hilt to provide rapid transit on *one route*."[11]

The financial experience with rail rapid transit adds to the evidence that rail facilities may be financially impractical even where the heavy initial capital costs have already been paid. The heaviest transit losses in Chicago have been in rail rapid transit operation. Losses from these operations have been offset by the financial returns from motor and trolley bus operations.

As urban population has scattered over a wider area with resultant low density of development, it has become increasingly difficult to support any form of mass transportation. But mass transportation by rail is at a disadvantage for several reasons. First it is confined to a specific route, so that origins and destinations off the route require feeder service of some kind. The bus does not suffer to the same degree from this disadvantage because it can use the entire network of highways to serve a wider area. Because of its relatively low capacity, the bus can often provide more satisfactory headways, and it requires only moderate density of population to find profitable loads.

The rail vehicle using an exclusive right-of-way encounters the greatest financial disadvantage in the fact that it cannot benefit from sharing facilities with other users as the bus can. Highways for buses are used jointly by automobiles and trucks, and the bus is therefore obliged to pay only part of the cost. This is of major importance in view of the fact that home-to-work travel is concentrated in only a few hours of the day and only five days of the week. Thus although it may be economically feasible to maintain and modernize commuter rail facilities that are already in place, it appears from an economic standpoint that the development of new transportation facilities favors the express highway and bus.

It has been suggested that rail rights-of-way could be reserved on expressways wherever high-density development produces fairly well-established patterns of heavy traffic

[11] Everett G. Crandon, "Let's Face Facts!" *Mass Transportation* (February 1954), p. 23.

movement. The Eisenhower Expressway in Chicago, for example, provides rail transit line in the median strip. Under most circumstances, however, the location of expressways at a considerable distance from passenger origins and destinations would require feeder service by bus or private passenger car. Thus the use of buses on the expressway in the first place would be more feasible.

Perhaps the most significant argument against rail rapid transit on an exclusive right-of-way is that the peak loads creating the principal passenger transportation problems in urban areas occur not only on weekdays but on weekends and holidays as well. Saturdays, Sundays, and holidays, nearly one third of the days of the year, generally create the heaviest travel peaks. Very little of this traffic is accommodated by local public carriers.[12] In providing transportation facilities for the metropolitan area, therefore, it is necessary to consider not only weekday needs, but social and recreational travel requirements on weekends and holidays as well. The question is whether the total job can be done with one transport system or whether it will be necessary to provide a system of rapid transit facilities and express highways too. From an economic standpoint a system of transportation that can serve at all times appears to be preferable.

But a further question with respect to the relative merits of rail and bus transportation is whether under existing patterns of residential development and the diffusion of trip origins and destinations, it would be possible to attract sufficient patronage to rail service. In New York, for example, traffic studies show that if rail facilities were effectively improved, the number of trans-Hudson passengers who would be attracted would be relatively small because a large percentage of mass travel has shifted from rail to bus for reasons that improved rail service could not alter.[13]

[12] In the New York area about one third of all Hudson River crossings occur on Saturdays, Sundays, and holidays.

[13] Port of New York Authority and Triborough Bridge and Tunnel Authority, *Joint Study of Arterial Facilities, New York— New Jersey Metropolitan Area* (1954), p. 9.

It has further been observed that little traffic could be expected to shift from automobile to rail because commutation by private auto in and out of Manhattan in the peak hours is negligible. Even if it were feasible to reduce by 50 per cent the number of passengers moving into Manhattan from New Jersey by automobile during rush hours, the effect would be a reduction of only 3.5 per cent in total Manhattan street usage.[14] Thus "it is by no means certain that the building of any rapid transit lines, even with a public subsidy, would necessarily redivert passengers from highways to a given set of rails and thereby eliminate the street congestion and highway problem."[15] The New York–New Jersey mass transportation problem is not how to accomplish a shift from auto use to public carriers, but how to provide better mass transportation service.

The New York situation is in many respects unique. In other urban areas the possibility that traffic congestion would be relieved by a shift of commuter traffic from private transportation to public carriers is substantial. The question is whether improvements in mass transportation can effect such a transfer, and if so, what types of facilities would be required and what costs would be involved. In San Francisco the study proposing rapid transit for the Bay Area concluded that a system of interurban rail service on exclusive right-of-way would offer a feasible solution to the commuter congestion that now stifles the highway and bridge approaches to the city.[16]

In reaching this conclusion, the position was taken that the bus could not provide the standards of mass transportation service necessary to compete successfully with the automobile because the absence of an exclusive right-of-way

[14] Information from Frank W. Herring, Director of Planning, Port of New York Authority.

[15] Walter Hedden, *A Review of the New York–New Jersey Metropolitan Rapid Transit Problem* (December 16, 1954), p. 8.

[16] Parsons, Brinckerhoff, Hall, and MacDonald, *Status Report to San Francisco Bay Area Rapid Transit Commission* (May 11, 1955).

subjects the public carrier to the same congestion that plagues the private automobile. The commuter caught in the traffic jam would prefer the comforts of his own vehicle to the added misery of a crowded bus.

The suggestion that an exclusive right-of-way is necessary to achieve a standard of mass transportation competitive with the automobile raises the question whether such a right-of-way needs to be a rail facility, which is subject to so many disadvantages. An exclusive right-of-way might be obtained by reserving special lanes for buses on expressways, or if new capacity needs to be provided, it would appear to be less costly to build duplicating highway facilities and possibly to permit their exclusive use by transit vehicles in peak hours. Highways, instead of being idle in off-peak hours and on non-working days as would rail rights-of-way, would be available to serve this off-peak traffic as well as commutation requirements. Thus express highways would be likely to attract sufficient total traffic to pay for themselves.

The situation is different, however, in the central core of the city or where rail rights-of-way are already available and the marginal cost of making such a system more attractive is less than the cost of alternative arrangements. Where facilities do not now exist and would have to be constructed, however, the contention that rail solutions can be adapted on any large scale to the accommodation of today's traffic patterns is dubious. In older communities that have grown up around mass transportation the use of rail rapid transit continues to be necessary and feasible; but even in these circumstances the trend in patronage is downward.

Possibilities of Mass Transport Innovations

Obsolete transit vehicles, poor ventilation, inadequate headways, inconvenient schedules, and overcrowding have added to the reasons for not using public transportation. Today the choice between public and private transportation is often a choice between a modern automobile and antiquated transit equipment. What are the possibilities, then, of providing good transit equipment and service that would provide a true comparison of the merits of the two methods?

The answer may be found in the development of new types of mass transport. Some indication of the outlook is suggested by answers to a questionnaire addressed to transit executives asking what urban transportation might be like in a typical big city in the year 2000.[17] The principal dilemma recognized by executives of the transit industry was whether the downtown area would still exist in the year 2000, but the conviction was expressed that if downtown should survive, the transit industry would confine its operation to that area, and automobiles would no longer furnish downtown service. The consensus was that streets in the future would make use of moving belts for pedestrian traffic, and that extensive perimeter parking would be provided for automobiles and trucks. According to one transit president, "no vehicle for individual transportation as we know it now will be permitted in the so-called downtown area."[18] Instead, the street or sidewalk level will be devoted to pedestrians exclusively, and on a second level transportation will be provided in the form of an endless belt or moving sidewalk reached by escalators.[19]

The belief was also expressed that most downtown traffic will be carried off the streets, either underground or overhead. Where rapid transit systems already exist, they will be improved and better equipment will be made available with great advances in automatic control. New off-street mass transit systems, however, may be of the conveyor type. "By substituting continuous conveyors for present mass transportation methods we eliminate the congestion resulting from batch loading, and the time lost in waiting for ve-

[17] *Mass Transportation* (January 1955), p. 40.
[18] The same.
[19] The chairman of one transit commission foresees lightweight jet-propelled roller skates for the shorter distances.

hicles."[20] For local circulation on the surface streets, small electric-powered vehicles that will eliminate noise and noxious fumes could be developed. Another technological possibility in urban transportation includes the development of automated highways which will provide electronic guidance and automated controls to prevent collision and at the same time increase speed and capacity.[21] Downtown congestion and the circulatory problem in large shopping centers has suggested the application of belt conveyor transportation to the movement of people. The first moving sidewalk or "speedwalk" was opened for passenger traffic in 1954 in Jersey City to carry commuters from the underground station of the Hudson and Manhattan Railroad up a 227-foot incline to the Erie Railroad Terminal. Other conveyor belts are being used at airport terminals. Additional applications have been explored for downtown department stores and office buildings, and for sidewalks from parking areas to large shopping centers.

In the coming decade the development of regional mass transportation by helicopter or convertiplane may provide the longer distance commuting services now provided by interurban buses and commuter rail lines. During 1963 over a million and a half miles were operated by U. S. helicopter lines carrying mail, express, and passengers in metropolitan service. Passengers are now being carried in the New York, Miami, San Francisco, and Los Angeles metropolitan areas. Most of this service was an extension of airline travel, covering the trip between airport and city center or between airports.

Scheduled helicopter service from downtown New York

[20] Sydney H. Bingham, *Mass Transportation* (January 1955), p. 35.

The same observer, however, "cannot visualize any major increase in transportation for suburban areas that is dependent on an exclusive and fixed right of way. People will travel between their homes and the urban periphery in their own vehicles or on some improved development of the bus."

[21] "Electronic Roads for Tomorrow's Traffic," *Business Week* (April 24, 1965).

to Kennedy Airport has cut peak-hour travel time from one hour to eight minutes. From the top of the Pan American Building at Grand Central Station, trip time to LaGuardia Airport is four minutes. Substantial time savings are also being accomplished by helicopter lines to twelve communities in the Los Angeles area. Altogether in 1964 a total of 606,000 passengers patronized helicopter services in the United States. Airport to city helicopter services are also being operated in England, Belgium, and the Soviet Union. In East Pakistan, 25-passenger helicopters provide service from the provincial capital of Dacca to sixteen cities within a 100-mile radius.

The size, speed, and economy of direct-lift aircraft have improved appreciably with the introduction of twin-jet power. Average seating capacity increased from 7 in 1957 to 20 in 1964. The next generation of jet helicopters may be the 60-passenger helibus, which could provide 20-minute service between New York and Philadelphia and comparable short-haul services for other pairs of nearby cities.[22]

The introduction of improved vertical and steep take-off aircraft will bring major changes in transport facilities for urban areas. It appears that air transportation will be able to compete effectively with surface carriers in the short-haul market, leading ultimately to a diversion of commuter railroad and suburban bus traffic to the air.

Review of the relative cost and service characteristics of public and private transportation indicates that regardless of technological change, travel requirements and consumer choice will continue to favor the automobile. Revision of public policy to provide equal treatment for all forms of transportation will not appreciably alter the patterns of movement that have developed to date. It can further be concluded that the competitive aspects of auto and transit

[22] Testimony of Stuart G. Tipton, Air Transport Association of America, before the Aviation Subcommittee of the U. S. Senate Commerce Committee, March 8–9, 1965.

have been overemphasized. Actually, their respective roles are to a large extent complementary. The urban transportation problem is in reality several different problems that vary in time and space. The advantages and disadvantages of public carrier and automobile transportation vary according to these differing circumstances.

In the downtown areas of most large cities, land use is so intensive that the attempt to accommodate private transportation under all circumstances is not feasible. There, in peak hours, the emphasis must continue to be on mass transport. In the less dense areas outside the central business district, however, residential and commercial uses must be accommodated by a combination of private and public transportation with increasing reliance on the automobile as incomes and automobile ownership increase. The problem again changes in character as the setting shifts to the relatively low densities of the expanding suburbs, where public transportation is least able to meet the local needs of a scattered population and where there is nearly complete dependence on the private car. Each one of these types of movement demands a different transportation solution, depending on the density of urban development.

In addition to these geographic factors, the urban transportation problem varies with the time and purpose of travel. On weekdays there are two distinct types of problems. One is the peak-hour problem, the other the very different problem of furnishing transportation off-peak in the middle of the day and at night. In addition there is the weekend and holiday peak, when recreation travel again alters the nature of urban travel and of the facilities required to provide it. Failure to distinguish these different aspects of the transportation problem has prevented a clearer understanding of the solutions required to provide satisfactory standards of service. And failure to plan and operate the transportation system as a unified whole has made impossible the most effective use of available transportation facilities to meet the variety of conditions imposed by urban transportation demands.

CHAPTER V

Transport Financing Policy

The search for better ways to provide urban transportation has given rise to a number of financial and organizational innovations. The establishment of toll authorities, transit authorities, port authorities, parking authorities, and a variety of intergovernmental agreements and commissions demonstrates the role that new methods of administration, finance, and operation play in making sound solutions possible.

Despite the efforts to date, however, progress toward resolving fundamental problems has not been impressive. There appear to be several reasons for this. First, current action to provide highways, parking, and transit is still being taken independently without any definite concept of the relative roles that can be effectively played by the automobile and public carriers in the future. It will be necessary to establish a clearer understanding of the extent to which the two forms of transportation are competitive, and the degree to which they should be designed to complement and supplement each other.

Second, current methods of finance are still inadequate to meet the extraordinary requirements of highways, terminal development, and transit modernization. Moreover, there is no uniform policy with respect to the financing of transportation facilities, so that differences in financial methods rather than in consumer transportation needs and desires, are determining the nature of transportation investments.

Expressway systems will continue to require heavy financial support. Yet, the anticipated growth of traffic in urban areas raises the question how long newly provided facilities will be able to cope with the demand. The costs of keeping up with the need for new capacity indicate that more ambitious arrangements may be necessary to provide a complete metropolitan system of expressways, and that in some cities supplementary rapid transit systems offer the only hope of lasting solutions.

Efforts in recent years to establish new financial and managerial arrangements for transit have eliminated some of the financial problems of the industry. Public ownership and the establishment of authorities reflect many of the same problems and the same need for better solutions that have prompted similar administrative innovations in the highway field. These new approaches have succeeded in scaling down inflated capital values, eliminating duplicate transit services, reducing taxes, providing new capital, and removing burdensome rate and service regulation. However, the financial outlook remains poor as transit costs continue to mount and patronage continues to decline.

A third reason for slow progress is the fact that current efforts are generally being undertaken without organizational arrangements to make possible the necessary geographical coverage and the necessary physical co-ordination of facilities and services. With a few exceptions each sector of the transportation system is being operated in a vacuum, and a partial approach is leading to only partial solutions.

In the highway field, intergovernmental agreements have to some degree halted the piecemeal approach resulting from the absence of metropolitan highway agencies and the dissipation of highway revenues among a multiplicity of governmental units. Joint governmental arrangements, however, have not eliminated duplication of highway organizations and overlapping of road functions. The compacts among existing road agencies fall short of providing a unified highway operation for the area. Differences of viewpoint among the several governments involved introduce problems that cannot always be resolved in the best interests of the metropolitan area as a whole.

The fourth weakness of the current approach to urban transportation problems is that most of the steps taken thus far have concentrated on the supply side of the problem. Little effort has been made to analyze the factors underlying transportation demand with a view to bringing supply and demand into better balance. Yet the fact that demand continues to outrun supply regardless of efforts to increase transport capacity indicates the need for directing more attention to this neglected side of the problem.

These considerations plus the outlook for continuing urbanization and growing densities of urban traffic raise serious questions about the ultimate effectiveness of current measures. The need for a different approach is strongly suggested if urban areas are to reverse the present trends toward an even more critical transportation problem in the years ahead.

The Debate over Subsidized Transport

The system of transportation that will be necessary to serve the urban area in the future will involve very large capital outlays. To date not even minimum transportation requirements have been met by available financial resources. Furthermore, there is no indication that the scale of construction called for by a bolder attack on urban congestion can be supported by conventional approaches. The question how the necessary funds are to be raised, therefore, is of governing importance in accomplishing a physically adequate transportation system.

But the problem of finance is not simply one of obtaining the necessary funds. Methods of financing transport services, and decisions governing the allocation of funds to various types of facilities, will determine to an important degree the total volume of traffic and its distribution among different transport methods. In other words, financial policy exerts an important influence on demand as well as on supply. The key issue in the evaluation of alternative solutions is self-support vs. subsidy. Relevant considerations include the financial position and policies of municipalities, the advantages of user charges, and the feasibility of self-support.

The cost and service competition between automobile and transit described in the previous chapter derived in part from the inherent characteristics of the two methods of transportation and in part from differences in policy affecting each method. Would more people use transit if the cost of public carrier and automobile transportation were on a more comparable basis? To what extent is automobile transportation being promoted by favored treatment on the part of government and to what extent have public policies affecting transit made it impossible to provide standards of service that the transit vehicle is capable of providing?

These questions are difficult to answer because transportation facilities have been financed by a variety of methods. In some cities transit service has been required to pay its way and contribute tax revenues to the city, while elsewhere it has been partially supported by general tax revenues. Highways have been paid for partly by the motorist through user taxes or tolls and partly by general tax funds. Parking facilities may be free, partially self-supporting, or provided at a profit. This variety of financial policies confers varying degrees of competitive advantage on one form of transportation or another.

The problem is illustrated by the competitive struggle between railroads and trucks. Railroad spokesmen inquiring into the cause of urban congestion conclude that "the principal cause is the method of financing the magnificent improvement in highways leading into the cities." According to this view, "vehicular traffic is enticed by hot house methods into the cities which become so congested that transaction of business in them gets more and more expensive."[1] The "hot house methods" refer to the belief that trucks are not paying a fair share of the highway bill in contrast to full private financing of railroad facilities.

Dissatisfaction with the equity of present financial methods has also been expressed in a report of the Mayor's Transit Study Committee in Milwaukee. The committee has pointed out that the city spends millions of dollars in build-

---

[1] "Editorial," *Railway Age* (March 16, 1953), p. 19.

ing expressways and maintaining streets, but that a considerable gap exists between the cost of providing these facilities and the amount of money realized from motor vehicle taxes. The city spent many more millions of dollars for highways than its receipts in motor vehicle fees and taxes. As a general rule it has been observed that in large cities part of the difficulties of congestion and finance lies in the underpricing of services in relation to costs. "That is, user charges do not cover direct and indirect cost of highway use, and consequently highway travel in high-density areas tends to be heavily subsidized."[2]

The matter of resolving the alleged inequities in urban transportation finance has given rise to two schools of thought. One takes the position that if each form of transportation paid for itself, the cause of unfair competitive advantage would thereby be removed. The other opposes specific use charges and earmarking of revenues, and holds that transportation should be financed, as most other public services, out of the general fund.

Adherents of the latter view state that since the municipality depends on good transportation for its vitality and accessibility, the cost of transportation facilities should be met through general taxes on the entire community. For example, one transit analyst has maintained that "the fares don't begin to pay the costs of a good transportation system, so it's really not necessary to collect them."[3] Less extreme and more typical is the view that acceptable standards of public transportation service cannot be obtained on a self-supporting basis, and that low fares subsidized by tax revenues will encourage transit patronage and more than compensate the taxpayer by maintaining urban property values.

The issues are illustrated by the controversy over New York subway fares and the transit deficit. For many years

[2] Lyle C. Fitch and Associates, *Urban Transportation and Public Policy,* Chandler Publishing Company, San Francisco, 1964, p. 129.

[3] Remarks by Steen E. Rasmussen reported in the Lincoln, Nebraska, *Journal* (February 24, 1954).

it was apparent that the continuing rise in operating deficits on the subway system posed a threat to municipal solvency. With a total deficit of over $100 million in 1953, New York State officials took the position that "New York will never be able to do anything fundamental about better schools, parks, roads and health services if it keeps sinking tax funds into the bottomless pit beneath its streets."[4]

The view of the city, on the other hand, was that higher fares to cope with the deficit would be self-defeating, because the increase would discourage riders and shift traffic instead to the already overcrowded streets. Moreover, higher fares would also drive business away from the city, thereby reducing downtown values and tax yields. According to some observers, "New York's prosperity is so intertwined with the availability of high-speed low-cost transportation that it is perfectly appropriate to assess part of the cost against industry and real estate."[5]

The solution of raising the fare to cover the mounting costs of transit had been tried in 1948 when the rate was doubled from five to ten cents. The increase resulted in a 13 per cent loss of patronage during the first year. Part of the loss was due to other causes that would have reduced mass transport patronage in any event, including the postwar return to the automobile and the long-run trend away from transit, which began in the 1930's.

Whatever the precise effect of the fare increase, the fact remains that 267 million riders deserted the subway after the increase went into effect, and it was feared that a further fare rise to 15 cents—while yielding higher revenues—would result in still lower patronage. Resulting business losses and depreciation of property values, it was contended, would shrink tax yields, so that failure to subsidize transit through business and property levies would prove to be false economy. Moreover, with higher subway fares a large number of patrons would elect to drive, thus adding to street congestion.

[4] A. H. Raskin, *New York Times* (March 5, 1953).
[5] The same.

New York State authorities were unmoved by these statistics. They continued to express concern over the deteriorating financial position of the city and insisted that only a further fare increase could stem the mounting deficit. In his special message to the legislature, the Governor of New York declared, "It is generally agreed that no long-range fiscal plan can be evolved for the city so long as the transit deficit remains part of the city budget."[6] The Governor cited the following reasons to support his contention that an authority should take over all city-owned transit facilities and operate them on a self-supporting basis.

1. It will remove the transit deficit from the city's expense budget and thereby avoid transit competing with schools, hospitals and other vital services for operating funds.
2. It will eventually release transit construction funds for extensions and improvements from the city's debt-incurring power and consequently enable greater use of such funds for building schools, hospitals and other essential public improvements.
3. Its independence and organizational set-up should lend itself to a more efficient operation, thereby reducing its costs and revenue needs.[7]

The Governor pointed to the fact that under city operation the subway had continuously deteriorated to the point of becoming "dirty, run down and overcrowded . . . yet there is no prospect of its improvement in the hands of a debt-ridden city with no money to improve service or buy badly needed equipment."[8] The message concluded that while the operating deficit should be met by a fare increase, "it seems inescapable that real estate must continue to carry the capital costs."

The view that transit costs must be subsidized was developed by the Mayor's Committee on Management Sur-

---

[6] Special Message from Governor Dewey to the New York State Legislature, March 10, 1953. *New York Times* (March 11, 1953).

[7] The same.

[8] The same.

vey. The committee found that the total cost of a subway ride, including interest and depreciation, was 16 cents, but that a self-supporting transit system would have required a fare of 18 cents to compensate for the estimated decline in traffic that would have resulted from the higher fare. Such a fare rise, in the view of the Mayor's Committee,

. . . would probably result in such reduction in the use of the transit facilities as to constitute a serious economic loss to the community, with an adverse effect upon riders as well as business concerns in the city . . . and the riders who would be hurt are not those with incomes over $100 a week, but those with incomes under $50 a week, particularly those with large families.

Furthermore, most of those who would stop riding because of a fare increase would not be the workers, who have to travel anyway, but the shoppers, customers and amusement patrons. This loss could be very serious to the businesses affected and thus in turn to the real estate, especially in the center of the city.[9]

The Mayor's Committee justified the subsidy on the grounds that "the benefits of such transportation go not alone to the rider or customer, but also to property value at the two ends of the journey." Elsewhere the opinion has been expressed that "the best way to allocate the resulting transportation subsidy tax burden is thus in proportion to the value of land."[10]

A similar case for property tax support of transit in Chicago has been argued on the grounds that rail rapid transit extensions had increased land values as much as 1,000 per cent in 25 years, so that rapid transit financing "might conceivably and properly be financed in part through assessments or taxes levied on benefited properties." Furthermore, while only part of the urban population uses public

[9] Report of the Mayor's Committee on Management Survey, *Modern Management for the City of New York,* Vol. 1 (1953), pp. 140, 161.

[10] Luther Gulick, "The Next Twenty-Five Years in Government in the New York Metropolitan Region," *Metropolis in the Making* (1955), p. 71.

transit regularly, everyone uses it occasionally, so that plant and equipment have to be available to meet this occasional demand. Because "this readiness-to-serve is of tangible value to all interests in any metropolitan area," it is logical to grant some type of subsidy to help meet these standby costs.[11]

The Case for Self-Support

Despite the arguments advanced in behalf of subsidies, there is a stronger case for developing a transportation system that is planned and operated on a self-supporting basis through direct charges on the users. First, self-supporting facilities provide a large measure of relief to local governments that are burdened by many financial obligations aside from that of providing transportation. Despite the great wealth of cities, many municipal governments are in a weak financial position. More and better public services are being demanded at a time when high prices have rendered traditional tax sources inadequate to meet requirements. Municipalities have felt the squeeze more than other levels of government because they are dependent to a large degree on property taxes that have been slow to respond to mounting government requirements.[12] "It is no exaggeration by now to say that New York City exists in a state of chronic bankruptcy. It is not that municipal bills are not paid . . . but that, facing the fiscal problem, budget makers have had to reduce expenditures until the municipal services reached an almost impossibly low level."[13]

[11] Charles E. DeLeuw, "Economic Aspects of Planning Rapid Transit in Metropolitan Areas," Paper presented at the Fifth Annual California Street and Highway Conference, Berkeley, California, February 5, 1953.

[12] Coleman Woodbury, "Local Government Finance," *The Future of Cities and Urban Redevelopment* (1953), p. 649.

[13] Rexford Guy Tugwell, "New York," *Great Cities of the World,* William A. Robson, ed. (1954), p. 417.

The second compelling argument favoring user charges is that general taxes levied by municipal governments can generally be escaped without difficulty because of the small geographical area involved. Many of those who work or do business in the city make use of municipal services, including transportation, but by living in the suburbs, they avoid paying a fair share of the costs. These conditions are a good indication that the achievement of adequate standards of transportation for the urban area will be accomplished more readily through a self-supporting system than through reliance on general tax support.

The precarious financial position of many large cities is illustrated by the situation in Boston. In the fifties total valuations of Boston real estate were approximately three quarters of the 1930 figure. This lowering of the tax base had occurred as municipal expenditures for greatly augmented public services had risen. The increase in the tax rate necessary to obtain sufficient revenues from the smaller tax base had in turn been responsible for reducing the attractiveness of downtown property and further depressing downtown assessed values.[14]

According to the Boston Municipal Research Bureau, there are three principal causes for the failure of the tax base to keep pace with the growth of Boston: the rapid enlargement of the tax-free property list, the disappearance of large amounts of assessed valuation to accommodate large-scale public projects, and the relatively small volume of new tax-paying construction. Using 100 as an index of the situation in 1930, the valuation of taxable real property in 1951 had fallen to 78, while the index for valuation of tax-exempt real estate had risen to 150 and the property tax levy had jumped to 162.[15]

Trends in urban land ownership present a growing financial problem for every large community. In Boston, for

[14] Boston Municipal Research Bureau, *What Is Happening to Boston's Tax Base?*, Bulletin 167 (December 27, 1951), p. 2.
[15] The same.

example, about two fifths of all the land area, exclusive
of streets, is exempt from taxes. Public agencies own three
quarters of the tax-free real estate. They include the city,
the federal government, and the Commonwealth. The city
is the largest land owner, and public housing projects represent the largest recent addition to the holdings of the
city.[16] The federal government is the second largest owner
of public property by reason of its military and naval
facilities, veterans' hospitals, and downtown office buildings. Property owned by the Commonwealth of Massachusetts includes that of the Massachusetts Bay Transit
Authority, the port authority, the airport, and the Mystic
River Bridge Authority. When the Metropolitan Transit
Authority was established in 1949 its properties in Boston,
assessed at $23 million, were exempted from local taxa-
tion. They account for a substantial part of the total increase in state-owned property. The Logan International
Airport, valued at $21 million, is another large addition
to the tax-exempt list.[17] Private holdings include universities, colleges, schools, churches, and benevolent institutions.

One of the most significant factors in the declining tax
base of the city is the liquidation of properties being absorbed by major highway projects. Land and buildings
taken for the Boston Central Artery were assessed at over
$7 million, and the total will be increased as the route is
extended. Land takings for the Mystic River Bridge alone
cost about a half million dollars in assessed valuations,
and one traffic improvement project involving the redesign
of an intersection took another $632,000 off the tax books.

[16] Payments in lieu of taxes partially reimburse the city for
tax losses on federal and state supported housing projects.

[17] Boston Municipal Research Bureau, *What Is Happening
to Boston's Tax Base?*, Bulletin 167. See also Boston Chamber
of Commerce, *Report of the Committee on Municipal Affairs:
An Analysis of the Financial Problems of the City Government
of Boston* (January 23, 1953).

Building permit records for the city revealed that, from the end of the war through 1951, approximately 59 per cent of all construction involved tax exempt properties not including federal and state construction for which city permits are not customarily required.[18]

Transport modernization creates a two-fold financial problem for every city. As large sums are withdrawn from the general fund for the improved transport facilities, the general fund is at the same time reduced by the withdrawal of land from the tax rolls to make the improvements possible. Many governmental services including schools, hospitals, and welfare services of all kinds cannot by their nature be financed on a user charge basis and must depend partly or wholly on general taxation. To these normal costs of government, there will be added very heavy outlays for urban redevelopment. These conditions cast considerable doubt concerning the wisdom of financing costly urban transportation facilities through general tax sources. On the contrary, whenever a direct charge can be made for public services in accordance with use, as in the case of transportation, there is compelling reason for doing so in order to make available general tax revenues for other public purposes not susceptible to direct payments.

A further advantage of the user charge is its ability to exert a positive influence over the character and amount of transportation service demanded. The influence of price on transportation demand is illustrated in the report of the Philadelphia City Planning Commission on financing the Delaware River Expressway. The commission pointed out that the imposition of a toll for use of the expressway would have the beneficial effect of reducing peak-hour volumes on the expressway as well as on the feeder streets that would have to absorb expressway traffic. It was estimated that trips on the expressway for distances of one to

[18] Boston Municipal Research Bureau, *What Is Happening to Boston's Tax Base?*, Bulletin 167, p. 4.

five miles would be reduced 50 to 95 per cent by a toll charge whereas little effect would be felt on trips of 10 to 15 miles. The over-all effect would be to flatten out excessive peaks within sections of the road with the heaviest travel.[19]

If public transportation were provided without charge, a transfer from private automobile to public carrier might reduce urban street congestion and the outlays necessary for expressway construction. But savings in street costs would have to be balanced against the outlays necessary to support the transit system. Some light is shed on what would be the effect of various levels and types of user charges by studies of the transit fare structure of New York. It has been estimated that if no fare were collected on the New York subway system, the number of users would be 2,207 million per year (at 1949–50 levels of costs and traffic) and operating losses would be $168 million. With a 20-cent fare the operating profit would be $91 million, and patronage would be about half what could be expected if transportation were free. With a system of zone fares a 1.5 cent per mile rate would yield profits of $18.5 million and attract 1,711 million riders whereas a rate of two cents per mile would yield nearly $50 million in profits and accommodate 1,504 million people.[20]

The effect of attempting to encourage transit riding by providing free service might be either to reduce the amount of public funds available for other governmental services or to increase general tax rates. As a third alternative, competition for limited tax dollars might result in a low priority for the transit system, and therefore lead to deteriorating standards of service.

[19] Philadelphia City Planning Commission, *Economic Feasibility of the Delaware Expressway* (January 1955), p. 56.

[20] William S. Vickrey, *The Revision of the Rapid Transit Fare Structure of the City of New York,* Technical Monograph 3, Mayor's Committee on Management Survey of the City of New York (February 1952), pp. 92–93.

Improving the Pricing Mechanism

User charges offer many opportunities to promote better transportation in cities, but to date the pricing mechanism has not been used effectively. In many cases application of the user charge has produced undesirable results
from the standpoint of both the industry and the user. A
flat fare on transit lines means that the rider in the downtown area who travels a few blocks and stands pays the
same fare as the commuter who boards at the beginning
of the line, gets a seat, and rides a long distance. There has
been a substantial loss of business as a result of flat fare
increases that treat short-haul and long-haul riders alike.
These losses have been primarily among the short-haul
riding group, because those who would pay five cents for
a one-mile ride are generally unwilling to pay two or three
times that amount. They have the alternative of walking,
driving their own cars, or taking a taxi. Yet loss of the
short-haul rider means increasing the average length of
ride per passenger, and with a flat fare this results in
sharply lowering average revenue per passenger mile.

In the highway field, the gasoline tax has proved to be
seriously deficient as a price for road use as it is collected
at a uniform rate throughout the state, regardless of the
cost of providing the specific facility being used. Thus a
vehicle may be moving on an urban expressway costing one
cent per vehicle mile to provide, but with a five-cent gasoline tax the revenue paid averages only one third of a cent
per vehicle mile. In addition, the gasoline tax has been
rendered deficient by inequities in the expenditure of the
revenues collected. This in turn has led to the charge, previously noted, that the motorist is not paying for urban
highways, to the competitive disadvantage of transit.

It has been seen that urban motorists pay large sums
in user taxes that are never spent for urban streets. The
urban motorist, rather than being subsidized, is subsidiz-
ing the rural road. Many states return to the cities amounts

of motor vehicle revenue that are far from commensurate with the revenue generated on city streets. The result is either that the automobile is not provided with the facilities that could be constructed if the funds were equitably distributed, or that the city is compelled to provide out of general revenues the highway plant already paid for but not realized by the motor vehicle user. Actually, on a national basis, urban traffic contributes more highway user tax revenue than is spent for city streets from all sources. This does not alter the fact, however, that the city fails to obtain the benefit of these funds and is required instead to provide the needed support out of local non-vehicle taxes.[21]

The pricing of parking facilities presents a further problem. Automobile transportation in cities benefits substantially from free parking, either on the street or in facilities provided by government or business. The importance of parking charges has been indicated in previous computations of automobile transportation cost with and without such charges. One means of eliminating part of the free parking that induces urban residents to travel by car rather than transit is to ban curb parking or raise curb meter charges. Overcoming the advantage of free parking made available by retail establishments attempting to compete with suburban shopping centers is a different problem. The fact that such inducements continue to be offered attests to the effectiveness of the subsidy as a means of luring the motorist into the city. The effect is partly to divert business from transit lines and partly to keep customers from defecting to suburban shopping centers.

From the standpoint of the community, the increased automobile traffic persuaded by free parking to use the city streets calls for measures to assure that the motorist pays the increased costs incurred to provide any additional highway capacity. The objective becomes one of preventing

[21] The city rather than the transit industry is the victim of the inequity, and transit together with the automobile stands to benefit from municipal efforts to compensate for the shortcomings of state fiscal policy.

free parking from shifting the burden of highway financing from the subsidized parker to the general taxpayer. But the pricing problem here is a problem of obtaining revenue, and is probably unimportant from the standpoint of presenting equitable competition. For there is no indication that an added payment for highways would retard automobile use appreciably and aid public transportation. Automobile ownership and use have continued to increase despite rising costs.

Present pricing policies are also open to question because they not only ignore the peak-load problem but often tend to make it worse. On the railroads, the commuter who travels in rush hours and is responsible for the costly peak capacity of the system pays a reduced fare that averages 2.02 cents per mile compared to 2.5 cents for coach passengers using other than commutation services.[22] The Port of New York Authority charges less at peak hours than in off-peak periods by offering low rates for bridge and tunnel travel to regular commuters who generally travel in rush hours. Over a quarter of the users of the Port Authority bridges and tunnels pay reduced rates through commutation and discount ticket programs. A commutation rate for the Hudson River crossings was instituted in 1950, and commutation tickets are used by a large proportion of weekday passenger car crossings.

Transportation price policies, then, indicate that insufficient emphasis has been given to cost considerations and to the possibilities of improving financial capabilities or of influencing transportation demand in the interest of more efficient service. It is likely that more attention directed to the user charges in the transportation field may reveal that this relatively unexplored area offers important possibilities for a more effective system of urban movement.

In order to maximize revenues and patronage of public carriers, fares should be related more closely to distance, as is the case with intercity carriers. An application of this

[22] Interstate Commerce Commission, *Transport Economics* (July 1955), p. 9.

principle would be the introduction of low fares for short rides to enable the public carrier to compete with the automobile in the shorter distances where the cost advantage of using an automobile is greatest. Transit fares might also be related more closely to costs so riders in the downtown area are not forced to support suburbanites, and profitless lines are not permitted to be a drain on the rest of the system.

With respect to peak-hour problems, the possibility of adjusting fares and tolls to achieve a better balance between traffic volumes and transportation capacities might offer a logical alternative to the present practice of low rates and fares during the hours of greatest movement. And the persistence of low railroad commutation rates in the face of mounting deficits might justify experimentation with rates that might permit better equipment and a higher class of service. The success of the toll road appears to justify a policy of adequate charges for better service.

The potentials of regulating traffic through the pricing of parking facilities are considerable. Patrons arriving after the morning peak and leaving before the evening rush might be encouraged by special rates to avoid using the streets during the heaviest traffic periods. Parking rates might be tailored to encourage group riding or combination auto-transit patronage if these objectives were found desirable. Special vehicle plates providing for toll-free use of expressways or permitting the use of specified expressway and parking facilities might prove feasible in urban areas. In places where a rapid turnover of parking space is desirable, hourly parking rates should increase with time rather than decline, and thereby place a premium on all-day storage. Curb parking is generally free or metered at the low rate of five or ten cents an hour, whereas rates for adjacent off-street parking may be many times as high. Yet at the curb the provision of parking space is likely to be more costly than off-street space. Where curb parking is permitted, therefore, more realistic rates should be charged.

A major impetus to financing needed urban highways has been the accelerated national program of road construction based on user revenues collected by the federal government. But the accomplishment of complete systems of expressways for the urban area will require much greater outlays. Since additional financing arrangements will be necessary in the larger metropolitan areas, the adaptation of the toll as a method of financing high-cost facilities could provide a significant supplementary source of revenue.

It may be argued that just as commuters by rail and bus pay a fare for transportation into the city, the cost of express highways used by automobiles should also be defrayed directly by the user. It should be possible to collect a toll when the motorist enters the expressway the same as a fare is paid when the transit rider enters the bus. The objection to stopping once at the toll gate may seem minor compared to stopping frequently at intersections on conventional streets that in the absence of extraordinary financial methods may be the only alternative.

But if the mechanical difficulties of toll collection can be overcome, the questions of public acceptability and financial feasibility remain. The belief prevails that a choice between toll facilities and toll-free facilities is a choice between paying and not paying. The latter choice, under this popular interpretation, is preferred. An added problem is that proposals for toll roads and bridges often run up against the difficulty that alternative toll-free facilities already available may create an awkward competitive situation.

One of the weaknesses of toll finance to date has been the individual project approach. This defeats the purpose of developing a system of facilities that together are self-supporting. The situation is best illustrated by a survey of arterial facilities in the New York area. A joint study by the Port of New York Authority and the Triborough Bridge and Tunnel Authority concluded that major bridge and tunnel projects needed to be undertaken in the area. But the highway connections to these proposed bridge projects, which would involve outlays of approximately $189 million, were not included in the toll financing.

Yet the substantial surpluses collected on the bridges and tunnels would be made possible because of the access afforded by the approach roads. The fact that tolls happened to be collected at the crossings would not signify that the crossings were self-supporting and the rest of the project was not. It is apparent that restricting toll financing to those facilities that of themselves will be self-supporting cannot provide a system of transportation because it denies support for the less profitable feeders that are essential to the main lines. A broader application of revenue bond financing would have to be tried if toll pricing were to be adopted on any large scale in financing of transportation facilities in cities.

The Feasibility of Self-Support

Thus far it has been pointed out that self-support for urban transportation offers a number of advantages. User charges relieve the city of major financial obligations and thus aid in the provision of other needed projects. At the same time pricing provides a productive and dependable source of funds with which to raise the level of transport services, and it promotes equity by establishing a more satisfactory relation among competing transport agencies and between the amount paid and the amount and character of the service rendered. Pricing can also be used to influence demand.

Of course the desirability of self-support does not establish its feasibility, and neither the past record of the transportation industry nor recent studies of urban transportation provide assurance that adequate facilities for urban mobility can pay for themselves. But it has been noted in previous chapters that many of the financial difficulties of the transportation industry might be removed, for in many cases they can be traced to shortcomings of management and public policy that could be remedied. Current difficulties appear in some instances to stem from a long history of underpricing transport services. Depressing transit fares has meant that funds have not been available for modernization, with a resulting decline in standards and patronage. But experience with toll highways indicates that high rates can result in growth rather than decline of patronage. Thus better pricing for better service may well be the key to a self-liquidating transportation system in the future.

But public policies that are putting the transit industry at an unfair financial disadvantage will also have to be changed if the financial position of public carriers is to be strengthened. The exemption of transit from franchise taxes and from responsibilities for street maintenance would lessen the burden on public carriers. Relieving the carriers of responsibility for providing reduced fares for school children would mean a further benefit. The relaxation or elimination of burdensome regulation would give management greater discretion in determining rates and route extensions and abandonments. Traffic regulations and enforcement designed to expedite transit operations would reduce operating costs as well as improve service.

Changes in public policy are necessary to provide the greater freedom to compete that should rightfully be granted the transit industry. But the industry is free now to do many of the things that could provide a sounder financial base for mass transportation. The development of off-peak operations through charter bus service has been demonstrated by many companies to be an important revenue producer. Successful attempts to utilize idle capacity in the off-peak periods can be a decisive factor in improving the financial position of the carriers. Express bus operations have proved capable of generating additional traffic in many cities, and reappraisal of transit operations has led to more economical routing, shifts from rail to bus, and greater emphasis on favorable public relations.

Regardless of possible changes in public policy, the fact remains that some transit systems today are showing satisfactory returns only because they are providing unsatisfactory service. Many are saddled with antiquated equipment and heavy debt largely because of past errors of public policy and transit management that should have been altered long ago. In these cases past indebtedness on subway and rapid transit would have to be eliminated to permit future self-support. Today's riders should not be obliged to pay for these past mistakes. Even with financial responsibilities limited to current operations and modernization, however, revisions in policy to date have proved inadequate to meet the need. It is doubtful that public carriers can furnish satisfactory service and at the same time meet the test of self-support under existing conditions.

The situation has been different in the case of highways. There appears to be no unwillingness on the part of the motor vehicle owner to pay for the facilities needed to provide better service. Getting the money allocated to the urban area where the needs are greatest has been the problem.

In the past the level of motor vehicle user charges was far from sufficient to pay for adequate urban highways. As in other areas of the transportation field, the price of the service failed to respond to rising costs or demands for better service. However, the trend has now changed. Expenditures for highway construction in the United States, expressed in constant dollars, have almost tripled since the 1930's. When these figures are related to the greatly expanded volume of traffic, it is found that highway construction expenditure per mile of travel in 1963 was 0.8 cents, double what it was in the late 1940's. (See Chart 17.)

The ratio of construction expenditures to gross national product was between 1.5 and 2 per cent in the thirties. Then following a sharp reduction in road building during the war years, the percentage rose again to 1.4 in 1963. (See Appendix Table 21.)

Despite the higher level of road construction, there is still heavy emphasis on intercity facilities and a neglect of the cities themselves. Under existing laws governing the distribution of motor vehicle revenues, it is generally impossible to obtain the necessary user support for urban expressways except by raising state-wide rates of taxation

TRAFFIC GROWTH AND HIGHWAY CONSTRUCTION EXPENDITURE

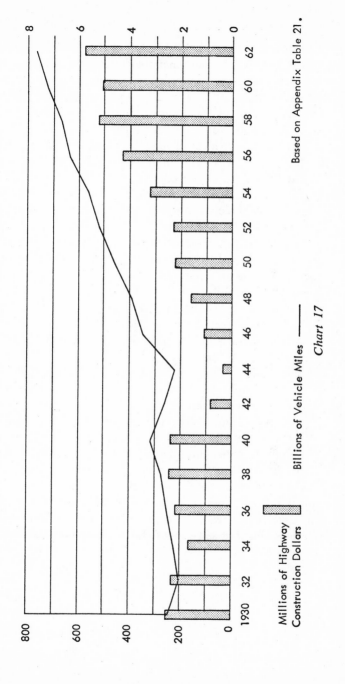

Millions of Highway
Construction Dollars

Billions of Vehicle Miles ——

Based on Appendix Table 21.

Chart 17

and dividing the proceeds among all the various rural and urban systems of roads that share the proceeds according to formula. In most states the feasibility of meeting urban needs in this way is limited by the fact that the over-all tax increase necessary to provide any appreciable addition to urban funds would be very high.

The annual payment of highway user taxes in the United States in 1964 was $10 billion. Of this amount 30 per cent found its way to city streets either in the form of grants-in-aid or actual expenditures by the state highway departments. This percentage is in contrast to the approximately 50 per cent of total traffic that is concentrated in urban places. Moreover, large cities with the greatest congestion generally receive the least user tax support in relation to the magnitude of their difficulties. Thus in many urban areas the possibilities of accomplishing programs of highway construction commensurate with the need continue to be thwarted by existing state laws.

It is concluded that present financial policies governing the provision of transportation in urban areas is in need of overhauling. There is no indication that present practices could support a truly modern system of urban mobility. But clearly the best chance of achieving the transportation needs of metropolitan areas in the future will be through a fully self-supporting system. The goal will require revision of state-aid policies, a closer relation between transportation costs and charges, a more effective use of the pricing mechanism, and new administrative arrangements in metropolitan areas. Financial success is not likely to be accomplished short of an integration of facilities and services aimed at self-support for the transportation system as a whole.

CHAPTER VI

Organizing Metropolitan Transportation

Efforts to solve the transportation problem in most metropolitan areas today have been complicated by the absence of unified administration. Responsibility for each form of transportation has been divided among different agencies within the central city and again among the various units of government sharing transportation responsibilities in the metropolitan area. The transportation problem is not being solved partly because no agency of government is responsible for its solution. Effective urban transportation cannot be provided without the organizational arrangements that make possible a unified approach to the planning, financing, and operation of the transportation system as a whole.

Divided Responsibility for Transportation

Methods of transportation that operate more or less independently in rural surroundings find that their fortunes are much more closely related in urban areas. There physical proximity and exchange of traffic create common problems of planning, administration, and finance. In most cities, however, each transportation facility continues to be managed separately and often without reference to the others.

The most obvious need in transport organization is to combine responsibility for automobile transportation and public carrier facilities, as already noted. Further need for co-ordination exists throughout the urban transport network. Traffic control measures may have an important effect on the efficient operation of mass transportation; low taxi fares may provide a substitute service for bus transportation at night; rail commuter services can be made more effective by integrating railroad and city transit lines; and the location and design of off-street parking facilities may have an important influence on highway congestion. Transportation policy decisions need to be determined with a view to their effects on the whole transportation system. The provision of each form of transportation separately means lost opportunity to achieve the best possible overall performance of the transportation system.

A number of proposals for urban transport administration have stressed the need for a unified approach. In St. Louis, for example, a single integrated agency was suggested, which would concern itself not only with the provision of transit but with traffic problems that are so important to the movement of public carriers using the street system. Transportation proposals for Chicago have suggested the consolidation of all municipal departments dealing with transportation matters, including the functions of bureaus dealing with streets, bridges, highways, tunnels, and airports. The municipal transportation authority recommended to accomplish this consolidation would ultimately be enlarged into a metropolitan transportation authority.[1]

The number and variety of public and private agencies responsible for transportation in a metropolitan area is illustrated by the situation in Philadelphia. The planning and provision of main highways is undertaken by the state highway departments of New Jersey and Pennsylvania, while other roads and streets are the responsibility of county and municipal governments and quasi-public agencies that include the New Jersey and Pennsylvania Turnpike Authorities, the Delaware River Port Authority, and the Delaware River Joint Toll Bridge Commission. This complex of organizations leads necessarily to a great deal of intercommunication and negotiation in an effort to achieve the obvious requirements for physical co-ordination.

[1] Ernest A. Grunsfeld and Louis Wirth, "A Plan for Metropolitan Chicago," *The Town Planning Review* (April 1954).

But no amount of time-consuming contacts on a voluntary basis can overcome the absence of a planned regional transport system. The highway departments on either side of the Delaware River proceed quite independently. Bus and truck terminals are provided by private enterprise, and automobile parking by both private and public jurisdiction. Mass transportation is provided by a private company whose operations include the use of city-owned rapid transit facilities. Jurisdiction over planning and financial matters is divided among the Philadelphia Transportation Company, the Department of Public Property, the City Planning Commission, and the State Public Utility Commission. Rail commuter services are provided by the Reading and Pennsylvania railroads, which in turn are subject to state and federal regulation. Under this administrative complex, the possibilities of achieving an interrelated system of metropolitan mobility are rather remote.[2]

In the New York area also, a variety of agencies operate in the transportation field. Interstate bridges and tunnels together with bus and truck terminals and airports are the responsibility of the Port of New York Authority. Water crossings that are not interstate are owned by the Triborough Bridge and Tunnel Authority, which operates seven bridges and two tunnels in addition to parking facilities and the East Side Airlines Terminal.[3] All other bridges are built and maintained by the city and administered by

[2] City of Philadelphia, Urban Traffic and Transportation Board, *Plan and Program 1955* (1956).

[3] The Triborough Bridge and Tunnel Authority is a consolidation of authorities created previously by the New York State Legislature including the New York City Parkway Authority created in 1938 and the New York City Tunnel Authority created in 1936. The New York City Parkway Authority was itself a combination of the Henry Hudson Parkway Authority and the Marine Parkway Authority, both of which were created in 1934. The Authority is administered by three nonsalaried commissioners appointed by the Mayor. The seven bridges operated by the Authority are: the Triborough Bridge, Bronx-Whitestone Bridge, Throgs Neck Bridge, Henry Hudson

the Department of Public Works. Parkways are built by the city and administered by the Park Department. Other highways are constructed by the city, some with state and federal aid. They are maintained generally by the borough presidents. Parking facilities are for the most part privately provided, but there is also a New York Parking Authority and some parking facilities are being provided by the Triborough and Port authorities. Mass transportation is provided by the New York City Transit Authority, private bus lines, and a number of private railroad commuter lines. The lack of unified action relates not only to planning, construction, finance, and traffic control but also to the fixing of fares, fees, and tolls.

The Mayor's Committee on Management Survey raised the question some years ago "whether the City of New York can deal with the transportation problem intelligently or effectively with such a disorganized approach." The committee stated: "With such an approach it is only natural for the various agencies which have a finger in the transportation pie to become competitive and to proceed without adequate integration with housing, schools, hospitals, recreation, and other transportation developments."[4] The same conclusions were expressed by the New Jersey and New York Transit commissions. "The time has come to deal comprehensively with the problem, and to develop a broad-gauged program for the co-ordination of transportation in the New York Metropolitan Area which will embrace all types of transportation, and which will lead to a better balance in the use of rail, bus, and auto transportation."[5]

Bridge, Verrazano-Narrows Bridge, Cross Bay Bridge, and Marine Parkway Bridge; and the two tunnels are: Queens Midtown Tunnel and Brooklyn-Battery Tunnel.

[4] New York City Report of the Mayor's Committee on Management Survey, *Modern Management for the City of New York,* Vol. 1 (1953), pp. 164–65.

[5] New York Metropolitan Transit Commission and New Jersey Metropolitan Transit Commission, *Joint Report on the Problem of Providing Improved Mass Transportation Between the*

In the metropolitan area of Washington, D.C., much the same situation prevails. The highway departments of the states and nearby counties of Maryland and Virginia are involved with the District of Columbia in arriving at physical and financial plans for highways and bridges. Differing views with respect to this aspect of the metropolitan transportation problem are complicated by wider differences in financial policies and capabilities, and by the fact that planning for this segment of the problem is entirely separate from the planning of mass transportation. Transit service is provided by a number of private companies that are themselves un-co-ordinated regarding service and rates, and they are subject to regulatory control by state utility commissions located in Baltimore and Richmond as well as the District of Columbia, and by the Interstate Commerce Commission. The National Capital Transportation Agency is responsible for the new rapid transit system.

Bringing together the various parts of the transportation system for purposes of planning, operation, and finance is not a panacea for the circulatory problems of the urban area. But without having at the outset the administrative machinery necessary to achieve a comprehensive physical plan, the basic tools with which to fashion a unified and effective metropolitan transportation system are missing. Decisions with respect to needed facilities cannot be wisely made unless alternatives can be weighed and the impact of action in one area measured in terms of its effect on another. Transportation solutions are limited when decisions are made separately for each segment of the transportation system. An over-all approach, on the other hand, makes possible physical and financial opportunities otherwise absent: the design of expressways to accommodate public transit; the construction of parking facilities in conjunction with expressways; the establishment of joint rates and services; selection of the best methods of movement

City of New York and New Jersey, Westchester, Long Island (March 3, 1954), p. 7.

to accomplish specific transportation requirements; and the pricing of services as an aid to achieving desired transportation objectives.

The Need for Regional Arrangements

In addition to the lack of machinery for relating the several aspects of the urban transportation problem, there are few agencies empowered to manage urban transportation on a sufficiently broad geographical basis to encompass the full dimensions of the problem. Modern highways and motor vehicles have erased political boundaries, yet numerous small units of government in the urban area still share the responsibility for providing transportation facilities. The continuing outward expansion of the city beyond its original municipal limits and the consequent growth of the area to be supplied with transportation services have made a regional approach increasingly necessary.

Obviously, the city is no longer in a position to handle the transport problems it has created. New York City, for example, contains some 315 square miles, but the total urbanized area of New York–Northeastern New Jersey is six times as large. (See Chart 18, p. 171.) In Boston, the metropolitan area is almost eleven times the area of the central city, and nearly two and a half times as many people live outside the city as in it. Only 34 per cent of the inhabitants of the Pittsburgh metropolitan area live in Pittsburgh itself, and the area of the metropolis is nearly ten times the area of the central city. And more than half the populations of the metropolitan areas of Los Angeles, Detroit, San Francisco, and St. Louis live outside the central cities. This tremendous growth of metropolitan population and area has not only magnified the physical problems of providing transportation facilities and services but has complicated these problems by rendering municipal governments obsolete as a means of supplying the transportation needs of the more expansive metropolis.

The result is that those who move to the suburbs con-

PROPORTION OF THE URBANIZED AREA INSIDE AND

OUTSIDE THE CENTRAL CITY

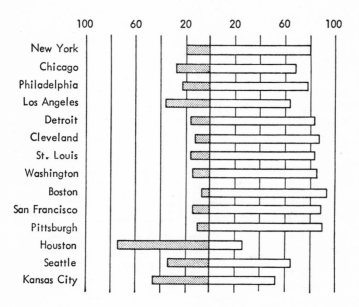

Area Inside Central City   Area Outside Central City

Per Cent

Based on Appendix Table 22.

Chart 18

tinue to contribute traffic to the central city as they return each day as commuters to make use of municipal streets and other facilities and services. Families that move out of the city are frequently replaced by lower-income groups who can afford no better than the undesirable neighborhoods left behind, and not only is their tax contribution of little help in supporting the continuing needs of the community but in many cases they add further to the cost of government by their dependence on public services.

Taxes paid by the central district of the big city must

carry the major burden of compensating for the public deficit elsewhere, and this in turn means high taxes on real estate and a powerful incentive for business and industry to join the movement of people to greener pastures beyond the limits of the city.

But an opposite effect also results from the separate jurisdictions of city and suburb. Often the central city attempts by tax concessions and liberal zoning provisions to retain or attract developments that will help to strengthen municipal finances. In many cases, however, this results in an undesirable density of development, so that the limited geographical expanse of the central city aggravates the trend toward greater congestion.

The need for regional government in the United States is by no means confined to transportation. Many services ought to be provided regionally. This need for additional services in outlying areas has in some cases overcome the desire of the suburbs to remain independent, and the advantages of controlling unplanned growth have often made annexation of fringe areas seem attractive to the central city. Thus one method of more readily providing such services is to move the city limits outward.

The possibility of annexation as a solution to the transportation problem is limited, however, and in any case, the desirability of this approach is questionable. Under existing "overly stringent, tortuous and antiquated state annexation laws," the obstacles to annexation have restricted the extent to which it has been attempted.[6] Individual annexations are generally small, and in already incorporated areas civic pride often makes it difficult to obtain the majority vote required to persuade established towns on the fringes to give up their independent status. The solution thus becomes feasible mainly in unincorporated areas rather than in groups of metropolitan towns where a unified administration may be most needed. Moreover, the possibilities of other arrangements for accomplishing the same

[6] John C. Bollens, "Metropolitan and Fringe Area Developments in 1953," *Municipal Year Book* (1954), pp. 41–52.

objective indicate that annexation is not necessary and may be an unsatisfactory solution. As one writer has stated: "For a number of years annexation took place because cities wished to boast about their territory. Most cities in America are now perfectly willing to let Los Angeles . . . have that honor." All of the advantages of annexation can be obtained without its disadvantages "by some form of central metropolitan control."[7]

Another approach to the metropolitan area problem has been the granting of extraterritorial powers through state legislation by which cities are authorized to extend their jurisdiction beyond municipal limits for certain purposes. Some cities go outside their boundaries to obtain an adequate water supply, and they may extend sewer facilities into the surrounding area to provide for adequate disposal. Approximately two thirds of the states have granted their cities such powers for gas and electric facilities, and thirty states have extended the jurisdiction of cities to enable them to control subdivisions in specified areas.[8] The use of extraterritorial powers as a means of accomplishing integrated urban government for the entire metropolitan area is also limited, however, and "suffers all of the procedural difficulties associated with annexation, with only limited compensating advantages."[9]

Another means of bringing the metropolitan area under a single local government is the use of the urban county as the administrative agency. The county is believed by some to be the unit around which to create a metropolitan government, granting that a "managerial" form of government must replace the traditional "magisterial" county before it can function successfully. There are now over 100 urban

[7] Walter H. Blucher, "The Problems of Our Metropolitan Areas," *The Journal,* American Chamber of Commerce Executives (April 1955), p. 5.

[8] Russel W. Maddox, "Cities Step Over the Line," *National Municipal Review* (February 1955), pp. 82–88.

[9] James C. Charlesworth, "Why Many Cities Are Too Small," *American City* (November 1954), p. 86.

counties that contain almost half the urban population of the United States. It has been suggested that more attention be given to developing the county into "the central agency of local government for some of the great complex urban areas," as this unit of government has shown a persistent vitality even in urban communities despite the often repeated assertion that it is doomed to disappear.[10] A "federal" form of metropolitan government could be built around the county and the municipalities within it, and short of an area-wide unit of government, an urban county is more likely than any other jurisdiction to include all or the major portion of the metropolis.

There are only a few cities in the United States in which city and county have been consolidated by special legislation, but again the use of this device to achieve urban regionalism is limited, as almost half the standard metropolitan areas fall within more than one county.[11] With the exception of Baton Rouge, this method has not been used in the last 50 years, and it is argued that the county is an "arbitrary and nonmeaningful unit of government" that should not be revived in any event.[12]

Regional Transportation Authorities

Regional authorities offer another alternative for organizing urban transport services more effectively. Theoretically they offer several advantages. One is the broader geographical coverage provided, another is the opportunity to deal with more than one form of transportation, and a third advantage is financial. The financial benefit derives from the fact that in the authority type of arrangement there is

[10] Victor Jones, "Urban-Counties—Suburban or Metropolitan Governments?", *Public Management* (May 1954), pp. 98–101.

[11] Betty Tableman, *Governmental Organization of Metropolitan Areas* (1951), p. 16.

[12] Charlesworth, "Why Many Cities Are Too Small," *American City*, p. 86.

freedom from constitutional debt limits, operations are exempt from most taxes, and the power of setting rates is generally provided without the intervention of an outside regulatory body. In practice, not all these potential advantages have been realized, however. The regional arrangements established to date deal with only part of the transportation problem and generally fail to cover the entire geographical area involved.

The operation of regional authorities is complicated by interstate problems. Thirty-two metropolitan areas in the United States extend across one or more state lines and many other metropolitan areas border on but do not cross a state line. Those areas that cross a state line contain 41 million people, one fifth of whom reside in a different state from the one in which the central city is located. In four of these areas more than 40 per cent of the population lives across the state line from the other 60 per cent. The five largest of these interstate metropolitan areas account for 30 million persons, or about one sixth of the total population of the United States. (See Appendix Table 23.)

Among the devices being used to co-ordinate policy in these areas is the interstate compact to establish agencies of joint administration. One of the most common uses of this arrangement has been to provide bridge and tunnel facilities. The Port of New York Authority, established by a compact between the states of New York and New Jersey, is the principal example of this approach to constructing and administering interstate transport facilities.

The area of responsibility of the Port of New York Authority is a region of approximately 1,500 square miles with its center at the tip of lower Manhattan. It covers the sections of New York and New Jersey most immediately affected by the harbor and its activities, and includes all or parts of 17 counties containing more than 200 cities and towns. The Authority operates six interstate bridges and tunnels, four airports and three heliports, six marine terminal areas, two union truck terminals, a truck terminal for rail freight, and a union bus terminal. In 1962, the operation of the Hudson and Manhattan Railroad, known as the Port Authority Trans-Hudson System, was also brought within the scope of the agency's responsibilities.[13]

Although the Authority is an instrumentality of the two states, a high degree of autonomy is essential as its functions of constructing and operating all kinds of transportation and terminal facilities would be difficult if concurrent action of the two legislatures were required. The states do exercise a degree of administrative control through appointment of the commissioners, the power of the governors to remove them on charges and after hearing, and through the veto power provided by law for the governors of each state.

The Authority has no taxing power and pays no state or local taxes on its property, no federal or state income taxes, nor state taxes on gasoline and motor vehicles. However, it does make payments in lieu of taxes on its terminal properties, equal in amount to the highest taxes collected on them before the Authority took over. Authority bonds are also tax exempt, and backed by the full faith and credit of the agency but not by the states.

Operations of the Port of New York Authority do not include responsibility for major highways, commuter railroads (except the Hudson and Manhattan), or transit. Inability to undertake projects that do not offer reasonable promise of eventual self-support has resulted in the neglect of many important aspects of the transportation problem in the New York area. Surplus revenues earned from motor vehicle traffic using bridge and tunnel crossings have been applied to airport and other projects only remotely related. The application of such revenues to other needs seems appropriate. If transit were to be included, however, the fear has been expressed that the financial position of the subway system in New York might jeopardize the whole financial structure of the Port Authority and close the door to further revenue bond financing. The fact remains that the financing of projects on an individual self-supporting basis in New York has failed to develop a net-

[13] The Port of New York Authority, *Annual Report 1963*, p. 1.

work of facilities capable of providing the best total circulatory system, and has made it impossible for the Authority to fill the need for a transportation agency capable of handling the total problem of the area.

An extension of the Port of New York Authority idea, and on paper the most comprehensive attack on regional transportation problems of an interstate character, is the Bi-State Development Agency created in 1949 by compact between the states of Missouri and Illinois. This agency has jurisdiction over two million people, more than 225 municipalities in the St. Louis metropolitan area, and 750 taxing bodies in six counties of the two states. The agency may "plan, construct, maintain, own and operate bridges, tunnels, airports and terminal facilities," and in addition may make plans for sewage and drainage facilities, co-ordination of streets, highways, parkways, parking areas, terminals, water supply, recreational and conservation facilities, land use patterns, and other matters in which joint or co-ordinated action of the communities within the areas will be generally beneficial.[14] Thus the St. Louis authority is designed to act as a general planning agency for the area with a scope of responsibility significantly broader than that of the Port of New York Authority. This type of organization attempts to avoid the limited approach to providing regional transportation services that typifies many authorities.

Most single-purpose transportation authorities have been established for financial reasons—because of the difficulty of providing a specific high-cost facility such as an airport or a bridge, or because the authority, by providing exemption from taxes and debt limits, can help to re-establish the financial position of bankrupt transit facilities. In all cases the immediate need for some particular facility or for financial relief in some part of the transportation system has resulted in adding a new administrative unit to the regional jig-saw puzzle.

Yet the query naturally arises whether further prolifera-

[14] The Bi-State Development Agency, *Annual Report, 1962–63*, p. 11; and *Annual Report, 1963–64*, St. Louis, Missouri.

tion of transport agencies is not defeating the objective of achieving improved transportation for the metropolitan area. Carving out more pieces of the problem for separate treatment aggravates one of the basic difficulties that needs to be overcome. If we continue simply to multiply authorities, we may be promoting, as one observer puts it, "a new type of government that may prove to be not only non-profit and nonpolitical, but nonsensical as well."[15]

The question is whether partial excursions into regional solutions can together meet the requirements for a total transportation system. If a single authority were to be established that would encompass all of the principal regional transportation functions, would this arrangement better satisfy the need for effective handling of metropolitan area transportation problems? A negative answer is indicated by the apprehension that a powerful transportation authority might fail to provide a satisfactory method of relating transportation to activities carried on by other governmental agencies. In a highly integrated metropolis, it has become clear that transportation plans have an increasing impact on other aspects of community development. Thus it may be unwise to establish a powerful public authority concerned with transportation alone without at the same time assuring a close relationship with other public agencies, including metropolitan area planning in general.

The ad hoc authority operating within a special district which is larger than the municipality attempts to solve the technical, administrative, and financial problems arising in connection with a particular service but isolating it from the complex of municipal services of which it forms a part. It may solve that problem, but only at the cost of weakening the general structure of local government in the great city and its environs, whereas the real need is to strengthen it.[16]

[15] Joseph E. McLean, "Use and Abuse of Authorities," *National Municipal Review* (October 1953), p. 444.

[16] William A. Robson, ed., *Great Cities of the World* (1954), p. 68.

Unless authorities are properly conceived, then, they may perpetuate the piecemeal approach to transportation in the metropolitan area. "The end objective is a unified government for an entire area to the greatest degree that such unification is politically desirable and economically feasible. Freezing a single function or activity into an authority may prevent the unification of all government."[17] Too often the authority is used "either as an 'easy' means of financing some new project, or as a substitute for an existing governmental unit which has failed to do a job satisfactorily."[18]

Despite these criticisms, the authority has proved capable of overcoming many problems of municipal governments. It can operate across municipal borders to supply common services to large areas and population groups; it has financed the construction and operation of highways, airports, parking facilities, and transit systems; it avoids the restrictions imposed by constitutional and statutory debt and tax limitations that are often as ineradicable as they are out of date; and it does not depend on the financial support of property taxes, which are neither equitable nor equal to the task. Financing improvements through user charges has proved feasible and acceptable by placing the burden directly on those who use the facility, and an authority because of its relative political freedom provides a degree of business management that is not found in ordinary municipal undertakings.

Role of the State

Another possibility in the transportation field would be for the state to fill the co-ordinating role sought through the creation of metropolitan or regional arrangements. An example of a multi-purpose state agency is the Massachusetts District Commission. The forerunner of this agency was the

[17] Carl H. Chatters, "Another Point of View," *American City* (February 1955), p. 116.

[18] T. V. Weintraub and J. D. Patterson, *The "Authority" in Pennsylvania: Pro and Con* (May 1949), p. ix.

Metropolitan Sewerage District in 1889 in which 33 cities and towns in the Boston area joined to provide a unified service. This was followed four years later by a Metropolitan Parks District comprising 38 cities and towns, and two years afterward by a Metropolitan Water District comprising 22 separate municipalities. By act of the legislature in 1901 the metropolitan districts were consolidated into one organization known as the Metropolitan District Commission, and today in connection with its park activities this organization is responsible for highways in the Boston metropolitan area serving the beaches and parks.

The Metropolitan District Commission is an agency of the Commonwealth reporting to the legislature and supported by legislative appropriations. All expenses are paid in the first instance by the Commonwealth and then assessed back on the municipalities for which the expenses were incurred. Construction of the parkways system was originally financed by bond issues amortized by assessments on the member communities, but recently these costs have been met from the Highway Fund by special appropriations for each project. State law now provides that the cost of maintaining boulevards plus Metropolitan District Police costs be paid from the State Highway Fund. Postwar bond issues for expressways in the metropolitan area have provided additional money for Metropolitan District Commission parkways.[19]

The Metropolitan District Commission, therefore, is a branch of the state government that integrates activities of a regional character that would otherwise be inefficiently provided by the many separate communities of the metropolitan area; and it provides a means of financing such regional services. But in the case of highways the functions of the commission have been limited, and no further transportation responsibilities have been assumed.

[19] Massachusetts Metropolitan District Commission, *Metropolitan District Commission, Development and Organization,* Office of the Secretary, March 1945, Revised September 1952, mimeo.

One shortcoming of the Metropolitan District Commission idea, however, is that a state agency with members appointed by the governor is unlikely to satisfy the requirements of stronger local self-government. According to one view, "on this ground alone we must reject such a solution of the problem of metropolitan government."[20] In any event, the District Commission is no longer responsible for metropolitan planning, and its failure to take on new responsibilities connected with the growing metropolitan transportation problem is evidence of a declining role for this agency in the Boston area.

Another approach to the metropolitan area problem was studied by the Boston Area Council, which is concerned with the achievement of co-operative planning and administration of throughways, circumferential highways, parkways, bypasses, subways, terminals, and other transportation facilities as well as water, sewage, drainage, and tax problems. The council, composed of delegates from each municipality involved, is dedicated to the idea that "suburbs should not be wiped out in a process of amalgamation to form a great city, but should be bound together in a federation—in this way providing for the orderly handling of common or joint affairs, while preserving inviolate the political integrity of each individual suburb." Objectives of the council are the preparation of a plan for the development of the community in the future and the preparation of material to convince citizens of the advantages of a metropolitan plan. "The time will eventually come . . . when the citizens are prepared to consider a comprehensive setting-in-order of metropolitan functions, and to establish an agency of government at the metropolitan level authorized to administer metropolitan affairs."[21]

It has already been noted that many states have aided in a variety of ways in bringing together the separate local governments in the urban area for expressway development

[20] Robson, *Great Cities of the World,* p. 67.
[21] William R. Greely, "The City or the Suburbs or Both?", *American City* (January 1955), pp. 83–85.

despite the fact that they have done "a better job of building highways than of solving the metropolitan governmental problems which the highways produced."[22] For example, a regional approach to the provision of highways has been partially realized in Los Angeles, Detroit, and Boston. But as long as the responsibility of the states is roads rather than transportation the results can not meet the increasingly complex requirements of metropolitan area mobility.

To correct this, the suggestion has been made that state departments of transportation should be created to absorb the highway departments and take on additional transport functions. But this does not localize the problem in the metropolitan area, and in most parts of the country today the hope that a state agency might be able or disposed to tackle the urban transportation problem comprehensively seems premature. Moreover, a state department of transportation could not take care of situations involving two or more states. For example, the Washington problem must be solved not only in the District of Columbia but in the states of Maryland and Virginia. The problems of New York relate not only to New York State but to New Jersey and Connecticut. On the other hand the problems of San Francisco and Los Angeles could be handled on a state basis. In any case, state governments may be expected to play a strategic role in providing solutions to the organizational problems of urban transportation as well as in promoting related regional planning functions.

Municipal Federations

Among the regional arrangements established to provide metropolitan transportation services, one of the most promising is the municipal federation. This is an association of local units of government that makes it possible for regional services to be supplied by a single agency on an area-wide

[22] Frederick L. Bird, "Metropolitan Area Finance," *American City* (January 1955), p. 143.

basis. Local functions continue to be provided by the localities. An illustration of this method is found in the metropolitan government of Toronto. During its rapid growth over the past 50 years, metropolitan Toronto became divided into 13 separate municipalities each geared to its own local pattern of development. None was very much concerned about what was happening to its neighbor or interested in the development of the area as a whole. Some units were able to finance needed municipal services and others were not.[23]

The answer to these problems was the establishment by the Ontario legislature of a system whereby the 13 municipalities may preserve their identity and continue to administer local services and at the same time combine for the provision of services that are metropolitan in nature and essential to the functioning of the whole area. The services for which the metropolitan corporation is responsible include water supply, sewage disposal, housing, education, arterial highways, metropolitan parks, certain welfare services, and over-all planning.

Under the Toronto Metropolitan Act of 1953, a Metropolitan Council was established composed of 25 members. Twelve represent the city and the 12 suburban municipalities are represented by the chairmen of their councils. The council was authorized to take over from the municipalities the facilities supplying regional services, and a Metropolitan Planning Board was established to prepare master plans covering land use, sanitation, parks, and transportation. The master plan becomes controlling on the municipalities, and they must conform to it in developing their own local plans.

Public transportation is separately provided. The Toronto Transportation Commission, which had been a separate authority for 30 years, was expanded into the Toronto Transit Commission with control over all public transportation in

[23] Frederick G. Gardiner, "The Municipality of Metropolitan Toronto—A New Answer to Metropolitan Area Problems," An address to the American Bar Association Convention, Boston, August 25, 1953.

the metropolitan area except railroads and taxis. The commission provides the nucleus of a rapid transit system composed of subways, surface lines and buses, and all independent bus lines operating in the suburbs became the property of the commission on July 1, 1954. The commission is required to make the system self-supporting through fares.

Before the new arrangement was established, no agreement could be reached on a co-operative basis on where the arterial highways of the 13 jurisdictions should be located and how they might be paid for. The Metropolitan Council has since then designated those highways in the area that should be metropolitan roads, and the Metropolitan Corporation has assumed all of the outstanding debt for such roads plus the costs of maintenance and extension. The corporation will also undertake the construction of new expressways, parkways, and other arterials to be paid for 50 per cent by the Metropolitan Corporation and 50 per cent by the Province of Ontario.

The whole of this metropolitan undertaking is being financed by a metropolitan budget with costs spread over the 13 municipalities in the ratio of their aggregate assessment. The Metropolitan Corporation issues tax bills at rates sufficient to provide the funds necessary for both current operations and capital expenditures.

The closest approach to this type of metropolitan government in the United States is the metropolitan government of Dade County, Florida.[24] Based on a study conducted by the Public Administration Service in co-operation with the University of Miami,[25] the new government relies on a strengthened county achieved through a municipal federation. The metropolitan government has assumed the functions of the present county as well as other activities considered to be more effectively performed on a regional

[24] The only other large area with metropolitan government in the United States is Nashville–Davidson County in Tennessee.

[25] Public Administration Service, *The Government of Metropolitan Miami* (1954).

basis. Each of the existing 26 municipalities in the area continues to provide purely local services. The metropolitan government provides local services to residents of unincorporated areas on the basis of service charges or through special service districts. Unincorporated areas are encouraged to incorporate or to seek annexation to nearby municipal units in order to obtain higher standards of service.

Metropolitan traffic and transportation are a responsibility of the new government, which is charged with assessment, collection, and distribution of property taxes; development and maintenance of arterial streets and major off-street parking facilities; development and operation of air, water, rail, and bus terminals; operation of public transit systems; development and supervision of urban renewal programs; operation of sewage disposal and water service; development and administration of a public education system; and over-all authority in the fields of planning, traffic engineering, health, welfare, and public safety.

According to the plan proposed, each municipality can supplement the level of metropolitan services in most of these areas. For example, in the field of planning and zoning the metropolitan government would make long-range land-use and zoning plans for the entire area, but each municipality could, by raising its standards above the minimum land-use plan, take over completely the zoning for that municipality. Thus no municipality is penalized in the matter of setting high standards, and no municipality is allowed to fall below the minimum land use plan created by the metropolitan government.

Before the Dade County metropolitan government was established in 1957, transit services were provided by seven private companies, with overlapping services, different fare structures, and an absence of express routes. The metropolitan Dade County Transit Authority now provides area-wide services on a self-support basis, having purchased the private bus lines through the issuance of revenue bonds.

Metropolitan government for the Miami areas has also led to the development of a long-range regional plan, including an over-all development plan for 1985. Close co-operation has been possible between planning functions and transportation agencies for the solution of immediate problems, and the opportunity has been afforded for more basic guidance of the growth of the area on a regional basis.[26]

In Philadelphia, the Urban Traffic and Transportation Board was appointed by the Mayor to study the transportation problems of the Philadelphia area and recommend a long-range program of transportation development. The board concluded that the interdependence of Philadelphia and the surrounding counties made necessary a regional approach to problems of mobility through an integrated regional transportation system. It was pointed out, however, that hundreds of governing bodies, boards, commissions, authorities, and private corporations providing transportation could not be expected to furnish the integrated facilities and services needed.

It was concluded that a regional transportation agency is essential to the provision of adequate transportation for the Philadelphia area, and that this agency should be designed with a view to the ultimate requirements of regional government. The agency would include eight counties in Pennsylvania and New Jersey, with the possibility of an interim solution limited to the five Pennsylvania counties involved. Functions of the regional agency would include planning of the transportation system and general control over regional highways, urban surface and rapid transit, suburban rail commuter services, taxis, and parking. Eventually, passenger and freight terminals for air, rail, marine, and highway transportation might be added. The agency would acquire direct management control of some parts of the system through contractual arrangements, such as with commuter railroads, or through purchase, as in the case of transit. In other areas, it would exert policy control through participation in the control of public expenditures and the determi-

[26] Institute of Public Administration, *Urban Transportation and Public Policy,* December 1961, Appendix Volume, pp. 90–93.

nation of standards of service and operation. It would construct such additions to the transportation system as might be necessary.

In operating or supervising the operation of facilities under its management control, the regional agency would be empowered to levy user charges, including fares, tolls, and fees, as well as excise or other taxes specifically related to the support of transportation facilities. Some of the income of the agency would be derived from federal, state, and local aid.

In order to assure the responsiveness of the regional transportation agency to the people, the governing council of the agency would have a majority of its members elected directly, while representation of the counties would be accomplished through selection by the governing bodies of the political units involved. An executive director would have full control over operating functions of the organization, while the council would be responsible for approval of the transportation plan, capital budgeting, and assistance in financing.[27]

Review of the arrangements thus far designed for the provision of transportation in metropolitan areas indicates a widespread search for administrative solutions. Thus far the quest has led for the most part to single-purpose agencies: the toll road authority, parking authority, transit authority, port authority, or airport authority. Some experiments and proposals have gone further to include several forms of transportation in a single agency and the pooling of financial resources. But thus far there has been limited success in putting into operation effective arrangements for regional transportation systems. The metropolitan area continues to improvise with outmoded administrative machinery that is unable to meet the pressing transportation needs of an automotive age.

[27] City of Philadelphia, Urban Traffic and Transportation Board, *Plan and Program 1955.*

CHAPTER VII

Transportation Demand and Community Planning

There is good evidence in the growing congestion of our cities that solution of the problem of transportation does not depend exclusively on the achievement of more and better methods of movement or on financial innovations and administrative reorganization. This is indicated by the fact that traffic congestion, which plagued the big city in earlier times, has continued and increased despite the fact that in recent years there has been more advancement in transport techniques than in all previous history. Solutions thus appear to depend not simply on measures designed to provide additional transportation capacity, but on the ability to develop urban communities in which satisfactory transportation is possible. The opposite course of attempting to supply transportation services to meet whatever demands arise from unplanned growth seems doomed to continuing failure.

The principal attack on the demand side must be through land-use planning and zoning, which establish the pattern of urban development and the location, size, and employment of urban structures, control the demands on public services, and underlie the generation of traffic. The present chapter will explore the possibilities of influencing the volume and character of urban traffic through the planning and redevelopment of cities and the introduction of measures designed to minimize unnecessary transportation.

Planless Growth and Traffic Volume

Concentration of population and vertical growth in downtown areas have created a demand for passenger and freight movement that has overtaxed the capacity of available transport facilities. These concentrations are being accommodated and promoted without regard to their effect on transportation demand. Resulting congestion appears inevitable as further growth takes place in already overcrowded areas. Arrangement of land uses may also be suspected of contributing to the problem by increasing the volume of crosshauling among functionally related but geographically separated activities, and especially by the separation of employment centers and housing.

The size of buildings and the intensity of land use made possible by the elevator have placed a heavy strain on the transportation system in cities. New York City has as many miles of elevator shafts as subway tracks. Eighteen million trips are taken daily on the 30,000 passenger elevators of the city, and 15,000 elevators are used for freight. Some people travel greater distances up and down in the course of a day than they travel on the ground. When they reach the ground all at once in rush hours, it is impossible for the horizontal transportation system to absorb the load with reasonable standards of efficiency.

Zoning has thus far had little effect on the race between those who supply transportation services and those who create the demand. Until the mid-1950's provisions of the Chicago zoning ordinance would have made it legal to house the entire population of the United States within the city limits.[1] The zoning in New York at one time was such that if development were provided up to the limits of what the ordinance allowed, building space would have accommodated 370 million people.[2]

In many cities too much land is used for buildings to permit satisfactory living or working conditions, let alone trans-

[1] Chicago City Council, Committee on Buildings and Zoning, *Proposed Comprehensive Amendment to the Chicago Zoning Ordinance, General Review* (January 1954), p. 3.

[2] *Plan for Rezoning the City of New York,* A Report Submitted to the City Planning Commission by Harrison, Ballard, and Allen (October 1950), p. 13.

portation. Insufficient space is devoted to parks and other open areas that lower densities and make the city more attractive. Congestion often results from the wasteful use of land supported by a pattern of streets and highways that usurps far too much space yet fails to provide mobility. In most cities one third of all the land is devoted to streets, railroad yards, terminals, airports, and parking facilities. Much of this space is required by urban street systems designed primarily for property access rather than for traffic. (See Table 25.)

TABLE 25. Major Urban Land Uses[a]
(48 U. S. Cities)

Type of Land Use	Per Cent
Streets and railroads	34.1
Single family residences	33.5
Public and semi-public	16.5
Industrial	6.5
Other dwelling	6.4
Commercial	3.0
Total	100.0

[a] Eldridge Lovelace, "Urban Land Use—1949," *Journal of the American Institute of Planners* (Summer 1949), p. 27.

In contrast to excessive density near the urban center, the sprawl of the new suburbs often results in densities that are too low to provide adequate transportation. There is often insufficient money to build the necessary mileage of modern highways and insufficient patronage to support needed public transportation. The result has been an absence of limited-access highways, growing traffic congestion, poor standards of transit service, and all the consequences of difficult commuting and inadequate community circulation.

Both over-concentration in urban centers and excessive sprawl in the suburbs have created conditions of blight and slums. These conditions have driven higher-income groups out of the densely populated areas. This has intensified the morning and evening waves of traffic between downtown employment centers and outlying suburban dormitories.

The reasons underlying the movement outward by those who can afford to do so is indicated by the large areas of blighted conditions in American cities. In Providence, Rhode Island, it has been estimated that 60,000 of the 70,-000 dwelling units in the city are located in areas that require some form of organized neighborhood improvement.[3] In Boston, close to 84 per cent of all dwellings in one redevelopment area were branded as substandard.[4] The result has been to focus on avenues of escape; to substitute transportation for satisfactory levels of living. The product is often a super-highway "jammed with people fleeing not from disaster but from the very city that is supposed to offer all the benefits that make life desirable."[5]

The disorderly flight to the suburbs, however, has been no answer to the problem. Suburban development has repeated most of the errors committed in the city. "The magic of a million homes a year still outweighs and obscures the frightful mess and sprawl these new homes are producing in and around our cities" and a new round of traffic congestion problems "that may well be insoluble."[6] The ability of the automobile to provide rapid movement from city to country is being nullified by the seemingly endless spread of urbanization that the automobile has promoted. The warning has already been sounded that by 1970 it will take an overnight trip to reach the countryside from the center of New York.[7] If we are to use more and more land for urban purposes, we must at the same time dedicate large areas of the city and its surroundings to parks and open spaces. This will meet the recreational and esthetic requirements of the

[3] Donald M. Graham and Robert F. Rowland, "Providence's Urban Renewal Program," *The American City* (September 1954), p. 118.

[4] Boston Municipal Research Bureau, *A Look at Redevelopment in Boston,* Bulletin 179 (November 16, 1953), p. 2.

[5] Lewis Mumford, "The Roaring Traffic's Boom," *The New Yorker* (April 16, 1955), p. 81.

[6] Albert Mayer, "Let's Plan Our Cities Before It's Too Late," *The Reporter* (November 18, 1954), pp. 35–36, 38–40.

[7] C. McKim Norton in *The Chicago Tribune* (June 13, 1955).

metropolitan area, and at the same time help to hold down gross densities.

It is apparent that solution of the urban transportation problem is closely related to both the restoration of close-in areas and the planning and development of better suburbs. Transportation solutions depend on community planning, and at the same time transportation developments can help to accomplish desired community goals.

The most important step toward solving the transportation problems of the city is to recognize the underlying conditions that have caused them. The St. Louis Plan Commission has concluded that if St. Louis does not provide an atmosphere for desirable living, it will continue to be abandoned by segments of the population it needs to retain. Consequently, plans for St. Louis call for lower densities that will provide "an attractive environment for living,"[8] which also means a reasonable environment for moving. In Boston, the City Planning Board points out that zoning is the basic tool through which the city can be designed for better living. "It is clear that an improvement in the quality of Boston as a place that is attractive for self-sustaining families to live is essential."[9]

The importance of transportation as a planning tool is stressed in plans recently outlined for Chicago, which recommended that a modern transportation system be constructed to help attack the unsightliness, inconvenience, and wastes now resulting from the high costs of urban living. Transportation is considered "the first item on the agenda and a prerequisite for the accomplishment of the other aspects of our proposal." It is pointed out that "without the reorganization and reconstruction of Chicago's transportation system the plans for industry, commerce, and the building of residential communities cannot be accomplished, congestion at the center will continue unabated, and Chi-

[8] St. Louis Plan Commission, *Comprehensive City Plan, St. Louis, Missouri* (1947), p. 10.

[9] Boston City Planning Board, *General Plan for Boston, Preliminary Report* (December 1950), p. 45.

cago will be a less prosperous and less livable city than otherwise would be the case."[10]

Looking at the impact of alternative transportation schemes, there is no question, for example, that the construction and full utilization of a subway system for rail transportation converging on the center of a city would greatly intensify land uses in the downtown area, whereas a beltline expressway circling the area would promote the objective of greater dispersal. In the San Francisco Bay area, a rapid transit system for interurban travel will more firmly establish the dominance of San Francisco itself and promote the concentration of population and economic activity. The development of expressways such as the New York Thruway or Route 128 around Boston supports and accelerates centrifugal forces, while the Central Artery in Boston may have the opposite tendency toward more intense downtown development.

The fact that transportation can achieve such opposite effects makes it a key to the urban future. But the key is not an open-sesame for the planner, because its use depends on what is technologically feasible and publicly acceptable. The best subway in the world built to serve an environment based on individual transportation will fail, just as the presence of the old circumferential highway around Boston never functioned as an incentive to disperse until a new road was designed to specifications for a community on wheels. Transportation has in general failed to operate effectively as a planning tool partly because too few communities have adopted plans to guide transport development and partly because legal, financial, and administrative obstacles have prevented effective use of transportation as a means of carrying out planning objectives.

In addition to the use of transportation as a tool for planning, there is need for planning that will recognize the relations between land use and traffic and thus the development

[10] Ernest A. Grunsfeld and Louis Wirth, "A Plan for Metropolitan Chicago," *The Town Planning Review* (April 1954), p. 14.

of communities in which effective transportation is possible.[11] The most obvious need is for achieving a balance between the amount of building space and the amount of circulation space. In older cities, the availability of modern transportation has made it possible for smaller buildings erected at an earlier time to be replaced by large modern office buildings, and this practice has made the downtown area increasingly congested. From a traffic standpoint, this congestion has been further accentuated by the multiplying use of the motor vehicle.

The relation between types of land use and traffic generation is shown in the accompanying table, which describes the situation in Detroit. For the area as a whole, it was found that residential land uses generate on an average weekday 29 person trips per acre compared to 269 trips for commercial uses, 37 trips for industrial activities, 33 for public buildings, and 3 for public open spaces. In nearly all distance zones from the center, it was found that commercial uses of land are by far the greatest traffic generators. In the zone 6 to 9 miles from the center, an acre of commercial land generates an average of 280 person trips per day compared to 42 trips per residential acre, 38 trips for industrial uses, 8 trips for public open space, and 46 trips for public buildings. (See Table 26.)

Planning and redevelopment along with zoning must create conditions that will permit closer physical relationship among land uses once thought to be incompatible. It should be possible to reduce distances that must be traveled, for example, from home to work. Zoning has always been predicated on the belief that certain uses of land should be separate because they are incompatible, and the virtues of zoning are often measured in terms of the degree to which zoning ordinances keep single houses, apartments, commercial, and industrial uses in separate areas. These concepts of the function of zoning have become obsolete as the possibility of constructing attractively designed and landscaped commercial and industrial establishments permits them to

[11] See Robert B. Mitchell and Chester Rapkin, *Urban Traffic, A Function of Land Use* (1954).

TABLE 26. Generation of Traffic by Land Uses in Detroit[a]

		Person Trips Per Acre by Land Use Type[b]					
Ring	Description	Residential	Commercial	Industrial	Public Open Space	Public Buildings	Total Land in Use
0	Core of Central Business District	733	1,797	153	—	945	1,522
1	Remainder of Central Business District	186	207	209	29	362	222
2	Remainder to three miles	65	194	92	10	89	74
3	3– 6 miles	56	218	48	3	26	58
4	6– 9 miles	42	280	38	8	46	50
5	9–12 miles	26	325	36	3	33	32
6	Over 12 miles	14	182	8	2	17	15
	Total Study Area	29	269	37	3	33	36

[a] *Detroit Metropolitan Area Traffic Study,* Pt. 1 (July 1955), p. 41.
[b] Excluding streets, alleys, and railroad rights-of-way.

be situated close to residential areas without detracting from them.[12] As the character of the factory has altered, and methods of residential planning have changed, the desirable approach to zoning has shifted from one of imposing regulations that separate different uses to the creation of conditions that will permit their coexistence. "A new type of zoning must be evolved which will permit, or rather insist, that the two get together, or the transportation problem will remain unsolvable."[13]

Poor planning and inadequate controls, then, are responsible for a large volume of transportation generated by the desire to escape the blight and slums of urban congestion for the more desirable living conditions of the suburbs. Yet the practice of taking advantage of today's mobility to escape the undesirable features of the close-in urban area will not solve the problem. The basic causes of congestion are

[12] Francis L. Hauser, *Title I Redevelopment the First Four Years and Defense Considerations,* Special Publication 8, National Association of Housing and Redevelopment Officials (June 1954).

[13] Henry S. Churchill, *The City Is the People* (1945), p. 93.

the absence of appropriate land use planning and the crowding of too many people into too little space. Until these problems are resolved, efforts to provide additional transportation capacity will be only partially successful. Failure to control the demand for transportation will continue to make chronic congestion inevitable. As one observer states: "The time has already come when we are wasting our substance by attempting to squeeze more cars, goods, and people into smaller and smaller areas. The simple geometry of the plan will surely defeat us no matter how long we postpone the day by ingenious engineering."[14]

Urban Planning and Renewal

In the central areas of large cities today, there is evidence of restoration and rebuilding on a scale that has not been equaled for many years. Whether the diseconomies of congestion being created by this rebuilding can be compensated by the economic advantages of central location is the question. The capacity of transportation facilities can be enlarged, and will be. But the effectiveness of these measures will continue to be canceled if increasing densities and planless arrangements of land use are permitted without regard for their effect on the movement of people and goods.

Comprehensive plans for community development and urban renewal are now going forward in many American cities. The nature and effectiveness of these plans will have an important influence on the size, density, and character of tomorrow's cities, hence on the nature of the transportation problem and its solution. Concern over the impacts of uncontrolled urban growth and resulting slums and blight has turned attention toward the necessity for planned communities designed to achieve predetermined objectives. This has led in turn to a growing awareness of the relations between

[14] G. Holmes Perkins, "The Regional City," *The Future of Cities and Urban Redevelopment,* Coleman Woodbury, ed. (1953), p. 39.

land-use patterns and the provision of community services, and the role that transportation can play in shaping the future city.

What density of development is to be sought for urban areas under various circumstances, however, is still the subject of considerable controversy. Although there are many social and economic reasons favoring a reduction of the congestion that burdens the close-in areas of most cities, there is considerable question whether the typical suburban sprawl of today is the desired alternative. Reduction of urban congestion is highly desirable in the interests of health, beauty, and convenience; yet low-density development has serious economic disadvantages. Suburban sprawl means higher costs for streets, electricity, water, gas, milk delivery, doctors, and every other service. "Every citizen on every private and business errand pays for low density of building with loss of time, fatigue or expense for transport."[15]

The problem, then, is to effect a compromise between horizontal and vertical growth. "If it is cheaper to extend a street into vacant land in the suburbs than to double-deck it, that is what will be done. Growth upward and outward will take place together, and each will be dictated by economic necessity."[16]

Review of comprehensive plans for cities in different parts of the world indicates growing recognition of the need for resolving these questions that impinge so closely on the relations between modern transportation techniques and land-use planning. Proposals are being made to limit the population of the city by zoning to assure a reasonable maximum. Allegiance to the "bigger the better" as a blueprint for municipal success is being supplanted by the concept that "effective planning of a metropolis is impossible unless a limit is placed on its maximum size and population."[17]

Planners are eyeing the possibility of reducing the spread of the urban area by preserving a buffer of undeveloped

[15] Kate L. Liepmann, *The Journey to Work* (1944), p. 103.
[16] Robert L. Duffus, *Mastering a Metropolis* (1930), p. 276.
[17] William A. Robson, ed., *Great Cities of the World* (1954), p. 103.

land around the metropolis and of directing further growth into smaller towns or new satellite communities. "The worst feature of many metropolitan areas today is that much of their environs consists of neither town nor country, but merely of land suffering from urban blight."[18]

Zoning provisions that will yield more open space and limit the height and bulk of buildings are being studied in relation to traffic generation, and many city plans call for the elimination of unnecessary street mileage to improve the circulatory system, to reduce the hazards and nuisance of motor traffic, and to make available for other uses the excess land now devoted to streets. The relocation of railroads and rail terminals is opening up large areas for more desirable development, and the problems of the pedestrian are being recognized in plans for shopping malls and the separation of foot traffic from motor vehicles.

More importance is now being attached to industrial location as the key to deconcentration and limitation of the size of the metropolis. In the Greater London Plan, it is stated that "the need for decentralization arises from the two-fold desire to improve housing conditions in those areas which are overcrowded, and to reduce the concentration of industry . . . which has made of Londoners a race of straphangers."[19] According to the plan, only by moving the factories out is it possible to achieve a closer relation between home, work, and community life and to reduce the time-consuming journey from place of residence to place of employment.

In decentralizing places of employment, however, the danger lies in destroying the rural surroundings and removing the central city still farther from the countryside to which it needs easy access. Thus a more recent survey of Britain's problems in the automotive age has concentrated on resolving the clash between urbanization and motoriza-

[18] William A. Robson, ed., *Great Cities of the World* (1954), p. 104.

[19] Patrick Abercrombie, *Greater London Plan 1944* (1945), p. 30.

tion by urban redevelopment aimed at securing both good access and good environment.[20] The Buchanan Report seeks a compromise between the two, which would involve different degrees of accommodation for automotive traffic depending on the size and function of the particular cities. But the principle is indicated by the fact that movement in any area would be limited to vehicles that belong there, and through traffic would not be allowed to intrude. Express highways and the development of functional areas would in combination help to minimize the conflict between urban living and the wheeled invasion of Britain which is threatening both town and country.

In the United States, the beginnings of comprehensive planning for the urban area are being paralleled by the redevelopment of slums and blighted areas and the renewal of neighborhoods threatened by blight. In 1964, a total of 777 cities had 1,634 federally assisted redevelopment projects under way or in the planning stage. These projects involved the acquisition and clearing of over 15,000 acres of slums and blighted areas. It is estimated that private and public construction in these renewal areas will total about $14 billion.[21]

All over the United States, the old urban centers of large cities are sprouting centers of restoration and new construction. These developments are intimately related to associated transport developments—new expressways, parking and terminal facilities, improved public transit and pedestrian malls.

In Rochester, New York, co-operation between private interests and public agencies led to creation of Midtown Plaza, which has revitalized a city core with a minimum of destruction of existing buildings. The area is ringed by an express highway and provided with three levels of under-

[20] *Traffic in Towns.* A study of the long-term problems of traffic in urban areas. Reports of the Steering group and Working group appointed by the Minister of Transport, London, 1963.

[21] National Planning Association, "Urban Renewal and Development," Planning Pamphlet No. 119, 1963, p. 10.

ground parking, financed by municipal bonds. Improved public transit by bus is also provided, including a new bus terminal for regional transport. The development contains office buildings, hotel, department stores and other retail establishments, a public auditorium, art gallery, restaurants, and air-conditioned covered pedestrian walkways. At the time Midtown Plaza was started, there had not been a major new private structure in downtown Rochester for twenty years. Now this development has triggered new construction in the surrounding area, including the $10 million Xerox Square adjoining the Plaza. At the same time, renewal in Rochester has created a new spirit of civic pride, greatly expanded community activities, and a powerful alliance of constructive public and private elements.[22]

Redevelopment for southwest Washington, D.C., required the clearance of several hundred acres of blighted land once accommodating 24,000 people. Approximately 80 per cent of the dwelling units in the area were found to be substandard. Included in the plan is the new South Mall overpassing and obliterating the railroad barrier separating the area from the rest of the city.

Another Washington redevelopment project, the Northwest renewal area, is the largest in the country. It consists of nearly a thousand acres and almost a quarter of a billion dollars worth of property. Its rehabilitation involves the razing of some 5,000 dwellings, about a fifth of the worst slum homes in the city. New street patterns and pedestrian ways, together with lower density of development and more open space, will make an important contribution to the solution of traffic problems.[23]

Redevelopment activity is often centered around the relocation of railroad facilities and the conversion of these sites to other purposes. Railroad terminals were located long before shifts in population and changes in the character and location of industry rendered them inadequate for

[22] Victor Gruen, *The Heart of Our Cities,* Simon and Schuster, New York, 1964, pp. 297–322.
[23] District of Columbia Redevelopment Land Agency, *Annual Report 1954,* p. 19.

the tasks of providing the city with needed transport services. Relocation of these facilities is providing substantial real estate capable of conversion to more appropriate uses and has at the same time removed one of the principal causes of downtown decay.[24]

In Philadelphia the new Penn Center, which was constructed on the site once occupied by railroad tracks and the Broad Street Station, has been made possible by the joint action of the city and the Pennsylvania Railroad. Plans include a twenty-story office building, a transportation center with airline and bus terminals, and a large hotel. A sunken pedestrian plaza with gardens one level below the street but open to the sky is planned to accommodate numerous shops and connect directly with the subway system. In addition, there will be underground connections with extensive parking garages, which in turn will connect with the expressway system.

Penn Center has a system for the distribution of goods to the shops and office buildings one level below the plaza. The site plan emphasized open spaces and improved traffic circulation through street closings and the widening of thoroughfares to and from the redevelopment projects and the major expressways.[25]

New Orleans established an authority to build a union terminal, and thereby eliminated five scattered stations. The result has been the removal of 144 grade crossings, speeding up traffic, and providing the city with "an entirely new transportation system."[26] This in turn has made possible many needed public improvements that otherwise could not have been undertaken.[27]

[24] Construction of Park Avenue over the railroad tracks of the New York Central in New York is an example of how valuable railroad property has been made to serve city needs.

[25] Philadelphia City Planning Commission, *Penn Center Redevelopment Area Plan* (August 1952).

[26] *Engineering-News Record* (May 13, 1954), p. 50.

[27] Properties to be abandoned when the New Haven Railroad carries out its plan for relocating Union Station in Providence, Rhode Island, will release a total of approximately 90 acres, part of which will be used to accomplish a 50 per cent increase

In Boston the 28-acre Back Bay Center was built on the site formerly occupied by the Boston and Albany railway yards. The Center includes office facilities and a retail center to serve 70,000 shoppers daily. There is underground garage space, hotel and convention hall, and connections with the Massachusetts Turnpike and the transit system. Of the people expected to use the Center when completed, it has been estimated that 36 per cent will come by car, 57 per cent by mass transportation, and 7 per cent on foot. Traffic will be in excess of volumes entering the business district in most cities of 80,000, but parking spaces will equal the number found in the central districts of most cities of 200,000.[28] Apartments, exhibit buildings, shops, and theaters accomplish a desirable mixture of day- and night-time demands for transportation.

This type of restoration on a large enough scale can mean a changing role for the central city in the years ahead. Residential and other appropriate land uses may be expected to displace present unplanned and in many cases irrational uses of close-in land. Urban redevelopment activities thus far undertaken have also involved a type of renewal that will not only alter the demand for transportation but will effect substantial alterations in transportation facilities themselves.

One of the major changes taking place is development of facilities for the pedestrian. Systems of walkways and greenways to insulate pedestrians from the interference of motor traffic are introducing a new human scale into the big city. Mobility for city dwellers depends a great deal on their ability to get about under their own power, and if cities can make walking more attractive, they can substantially reduce the demand for transit and vehicular movement.

The extent to which the pedestrian has been forgotten

---

in the size of the central business district. Providence City Plan Commission, *Railroad Relocation,* Publication 11 (September 1953).

[28] "The Back Bay Center," *Architectural Forum* (November 1953), pp. 105–7.

is indicated by the fact that in most cities traffic signals are geared exclusively to the needs of motor traffic, and no allowance is made to permit the traveler on foot to cross from curb to curb without interference. Yet many cities are coming to the realization that "there should be a pedestrian system of communications as efficient as that for the motor, and the less these two means of locomotion are provided in contiguity, the better for both."[29] The problem of safe walking in downtown areas will not in fact be solved until there has been a complete separation of pedestrian from vehicular traffic in areas of high traffic volume.

Pedestrianism may prove to be one of the significant transport achievements of current urban design trends. Both the outlying shopping center and redesigned commercial areas downtown rely on the restoration of foot traffic to make a more pleasant environment. In California, where there is nearly one vehicle for every two people, legislation has been enacted which recognizes that in certain urban areas, and particularly in shopping areas, there is need to separate pedestrians from vehicles "to protect the public safety." According to the law, this objective can be achieved by the establishment of pedestrian malls. Getting the city back on its feet may be as simple as that.[30]

[29] Abercrombie, *Greater London Plan 1944*, p. 11.

There is another incentive for encouraging foot traffic. A Harvard nutrition researcher has urged that the average chairbound worker should park at least four blocks from his office if he wishes to keep fit, but data on the average radius of pedestrian activity indicate that few motorists are disposed to walk four blocks. In Chicago, 50 per cent of motorists patronizing central area parking facilities walk a distance of one block or less to their destinations; 40 per cent walk from one to three blocks; and only 10 per cent walk more than three blocks. Studies reveal that it is difficult to attract patrons to facilities more than three blocks from destinations regardless of variations in rates and service. The Chicago Association of Commerce and Industry, *Parking Plan for the Central Area of Chicago* (December 1949), pp. 17–18.

[30] See Gruen, *The Heart of Our Cities*, p. 250.

Staggering Hours of Work

Another method of reducing central city transport demands is the distribution of the traffic load more evenly over the day. War conditions during the period 1941–45 demanded the staggering of work hours to achieve a better utilization of transit equipment, and 70 cities eased their public transportation problems considerably by this method. Staggering of work hours not only stretched the limited supply of transit vehicles but also helped to increase the effective capacity of street systems. Since the war, however, objections presumed to have come from both workers and business concerns have led to the termination of most staggered-hour plans, and the traffic problem in cities is more than ever a problem of everybody going to the same places at the same times. Urban transportation facilities are deluged with traffic for two hours in the morning and perhaps less time in the evening, so that many costly expressways and transit facilities must either be designed for the very high capacity demanded at the peak or built to more economical standards that mean acute congestion at the times that most people want to use them.

Many objections have been expressed against staggered hours: family life and social responsibilities are disrupted when friends or different members of the same family go to work at different times; people do not like to go to work early, or get home late; people in similar lines of business need to be at work at the same time in order to facilitate contacts. What may be the real feelings of employees and employers on the subject of arrival and departure times remains largely a matter of opinion, however, and the need for a thorough study of the economic and social aspects of the problem of staggered hours is apparent. No single step appears to offer greater potential for easing the urban traffic jam than a scientific attempt to modify the herd instinct that dictates traffic demands in urban areas.

The potentials of a comprehensive program of staggered work hours for metropolitan areas have been greatly increased by the outlook for a shorter work week. In looking to the future of the urban area "the dominant physical fact in the next quarter-century will be technological progress unprecedented in kind and in volume."[31] Routine office work now performed by large numbers of clerical help is being mechanized at an increasing rate, and in the future there may be a much lower density of workers per building for banking, insurance, and other large-scale office work. According to the director of the Bureau of Standards, "we are most certainly on the threshold of a business office revolution which will free the white collar worker from routine mental drudgery much as the industrial revolution of the last century freed the manual laborer from much physical drudgery."[32] The result of the revolution will be fewer hours in the office and at the factory.

There has been a continuing decline in hours worked per week, from 60 hours in 1900 to 38 hours or less today. The decline has recently been at the rate of about three hours per decade. A week of 32 hours may become general in non-farm activities within the next two decades, but in many industries the realization will be sooner. Hours of work will depend, however, on the way the worker elects to receive the rewards of increased productivity: whether in more income or more leisure. In the past quarter century, he has taken 60 per cent of his increased product in income and 40 per cent in time.[33]

A four-day week could come even before a 32-hour week if employees working 36 hours a week are given the choice of working nine hours a day for four days rather than spreading the work over five days. The four-day week may be favored over a five-day week with fewer hours per

[31] David Sarnoff, "The Fabulous Future," *Fortune* (January 1955), p. 82.

[32] Dr. Allen V. Astin, as quoted by Jerry and Electa T. Klutz, "Electronics Promise: Better Government for Less Cost," *Nation's Business* (February 1955), p. 41.

[33] Daniel Seligman, "The Four Day Week: How Soon?", *Fortune* (July 1954), p. 81.

day to avoid the extra day of commuting. Realization of either a shorter working day or a work week comprising fewer days could have a significant impact on the transportation problem and its solution. The four-day week, for example, might increase the peak load problem of the transit industry by compressing five days of business into four, and leaving the industry with three off-peak days.

If the five-day week were continued with shorter hours, a greater opportunity would be offered to stagger hours of arrival and departure. Under present hours of work the possibilities of staggering over any considerable time leads to very early arrivals or late departures. A reduction in the number of hours on the job would eliminate this situation. With automation increasing production and further reducing the length of the work week, staggering work hours over a six-hour or shorter day might permit spreading of commuter traffic around the clock, with business hours ranging from seven o'clock in the morning to noon, and closings staggered from one o'clock to six in the afternoon.

New Towns

It will not be enough to restore and improve the urban plant that has fallen into disuse and disrepair. In the next 16 years the United States will be building the equivalent of 1,000 cities of 50,000 population.[34] Much of the urban environment of the future has yet to be constructed, and in many parts of the country the work is under way on entirely new rural sites. As of 1964, there were at least 75 completely planned communities of 1,000 or more acres where developers were creating facilities to house more than six million people by 1980.[35]

[34] Robert C. Weaver, "The Significance of Public Service in American Society," 50th Anniversary Celebration, Institute of Public Administration, University of Michigan, May 25, 1964, p. 5.

[35] "New Towns for America," *House and Home* (February 1964), p. 123.

One such community on the edge of metropolitan Washington, D.C., is the new town of Reston, located on the highway to Dulles International Airport. This town provides a wide variety of housing, recreation, and community facilities, with a 900-acre area for industry and government use. There are single-family houses, town houses, and rental units in low-, medium-, and high-rise buildings, all with close access to views of open land. Recreational facilities include golf courses, two large lakes, stables, indoor and outdoor riding rings, and camping and gardening sites. Tennis and swimming clubs are provided in all seven villages making up the town, linked together with parks, lakes, town centers, and industrial areas.

The new community of Columbia in Howard County, Maryland, close to Washington, D.C., will house 150,000. Development of this suburban city includes provision for industrial and commercial employment opportunities to provide variety and vitality for an interesting and satisfying urban life. Here, too, new architectural design and land-use planning will be stressed to assure esthetic quality, good public services, open space, and convenient recreation and transportation.[36]

Another successful development of a new community is Conejo Village built 40 miles northwest of downtown Los Angeles. When the community comprising 11,000 acres was first being planned, the developer set aside 1,000 acres for industry and for green spaces, bridle trails, golf links, and other recreational facilities. Since then the town has attracted a large defense plant, research facilities, and light industries employing 6,000 town residents. According to a Stanford Research Institute report, there will be over 180,000 people living in the area by 1970.

Anticipated increase in urban traffic cannot be superimposed on existing congestion without causing a paralyz-

[36] Robert C. Weaver, "Urbanization in the Middle and Late 1960's," The Lorado Taft Lecture, University of Illinois, March 18, 1964, p. 9.

ing transportation crisis. Urban planning and development must bring about a dispersion of population and economic activity and a rearrangement of land uses that can reduce the waves of commuter traffic between suburb and central city to a more balanced regional flow.

The city we seek, as Clarence Stein first named it, is the regional city—a whole region that includes the countryside as well as the more densely built-up urban areas. The farmer and the city man, who have been separated through all history, can finally be brought together in the regional city. There the rural resident can seek the same benefits of education, medicine, and cultural activities as the city dweller. Conversely, the city man can enjoy the advantages of the countryside.

To realize this ideal of the regional city, the lesson that Lewis Mumford stresses is that urban renewal without reference to the suburbs, and continued suburban scatter without reference to the problems of the central city, are both of them ruinous processes. A balanced development will have to be achieved for the whole region. Such a development means that we must protect good agricultural lands, preserve landscapes for their beauty, and permanently surround built-up areas with green belts of farmland or woodland. The task is not one for planners alone, or for government alone. Industry, business enterprise, and cultural institutions of all kinds will have to take part in the replanning and reorganization of the new city. The result will be a kind of decentralization that will consist of many relatively small communities joined in a much larger urban association, where all the advantages of city and country will be able to be shared with the help of modern transportation.

If this concept of the urban America of tomorrow is to be brought into being, and if we are to reject both the over-concentration of the old-fashioned city and the endless monotony of megalopolis, the necessary steps are clear. They lie in new governmental arrangements—federal, state, and metropolitan—in new approaches to taxation and finance, in expanded programs for housing, renewal, and other community facilities, and in new planning concepts that relate official plans to the many actions taken in the private sphere. Among the most important tools for building the new urban community will be transportation. It is our new mobility most of all that offers the promise of the new regional city.[37]

[37] Wilfred Owen, *Cities in the Motor Age,* The Viking Press, New York, 1959, pp. 76–77.

CHAPTER VIII

Facing the Transportation Problem

America's cities and their sprawling suburbs have been unable to adapt effectively to the revolutionary changes in transportation that have made us a nation on wheels. Expansion of automotive transportation has outstripped the provision of highway and terminal facilities and created conditions of congestion that have become a serious threat to the urban economy. At the same time public transportation has suffered from declining patronage and changing conditions of use that have resulted in financial embarrassment and deteriorating standards of service. Along with this failure of the urban transportation system to meet the needs of modern communities, the possibilities of influencing transportation demand to bring it more nearly into balance with the supply of facilities have been almost completely overlooked. As a result we find the underlying causes of congestion multiplying more rapidly than measures for the relief of congestion can be applied. It is clear that a new approach will be necessary if there is to be any real progress toward solution.

The Present Drift

Moving in the morning and evening rush hours has been found to be the most critical transportation problem in American communities. It has been noted that the great surge of commuter movement between city and suburb has dictated the extraordinary physical requirements of the transportation system and imposed on the urban resident some of the most exasperating conditions of urban life.

For millions of families the principal rush-hour transportation problem is transit. Antiquated equipment, over-crowded vehicles, and slow service are the common complaints of the rider. Rising costs, traffic congestion, and a continuing loss of business are the common burdens of the industry. Patronage is down to the levels of nearly a half century ago despite the tremendous growth of population and industrial employment in the urban communities of the nation.

The prospects for radical improvement in the current situation—either for the transit passenger or the transit operator—are not promising. Many transit lines have ceased operations altogether; others are on the verge of bankruptcy. Where the financial position of transit is good, it is generally because standards of service are poor. The difficulty of handling commuter peak loads makes adequate service virtually impossible. The task of providing mass transportation that is satisfactory to the user and at the same time profitable to the company poses a major conflict that present policies have tended to accentuate rather than to solve.

The key to the troubles of transit, as described in Chapter III, is the progressive worsening of peak-hour problems. Heavy peak loads in the morning and evening rush have remained close to the prewar level, and loss of business has been concentrated in off-peak hours. This means that the decline in transit business has not permitted parallel reductions in cost. Capacity that is idle most of the day must be retained to serve the peak. The resulting feast or famine that typifies transit operations threatens the industry with bankruptcy.

In addition to off-peak hours, public transportation must cope with off-peak days. The postwar period that brought a five-day week means that business is now bad for transit on Saturdays as well as Sundays and holidays. On these days, as in the evenings, transit is being deserted for the automobile, which provides a superior service for the social and recreational trips that, aside from the journey to work, make up the bulk of urban travel.

Along with unprofitable hours and unprofitable days,

public carriers also have the problem of profitless routes. Public pressure has forced transit companies to serve the new suburbs where population is spread so thin that on many routes business is too light to pay operating costs. But public pressure demands that the service be provided. The fundamental conflict is between the interest of the carriers in economic survival and the interest of the community in the availability of service.

Rapid transit and commuter railroads have been especially vulnerable to the problems of the mass transportation industry and its patrons. Neither rail lines nor terminals are located where people live or work. While the motor vehicle can take to the road to follow the migration to the suburbs, transportation by rail has been unable to cover the vast new areas of urban growth that have filled in the spaces between the steel radials of an earlier age.

In the face of these difficulties the inclination of the urban resident has been to escape by automobile; but the attempt has met with less than complete success. The combination of transit difficulties and the shortcomings of highway and parking policies has created a serious dilemma for the American community. The urban resident who seeks relief by shifting from unsatisfactory mass transportation to the automobile is confronted by the equally unsatisfactory alternative of trying to drive on congested highways. Loss of patronage has added to the disabilities of transit, while increased auto use has compounded the frustrations of driving.

The dangerous implications of these trends for the future of the city have inspired no remedial actions commensurate with the problem. It has been shown that revisions in public policy may ease the burden on mass transportation, that transit management is in many instances making progress toward better service, and that expressway developments are providing relief in many areas. But there is little evidence of the all-out attack on the transportation system that could reverse the drift toward transport deterioration and the ultimate crippling of the urban economy.

Future Requirements

In most urbanized areas, the task of moving people is the problem of adapting to a motorized economy. This involves streets, expressways, parking, bus lines, taxis, terminals, and traffic engineering. In some of the largest urban centers, there are additional problems of railroad commutation and rapid transit. But in general the problem is motor transportation, and increasingly automobile rather than public carrier transportation. Neither economic analysis nor transportation history suggests a return to public transportation on a scale that would be decisive. Trends in automobile ownership show no sign of being reversed to conserve the space that motor transport has made plentiful. On the contrary, increasing leisure time, the spread of the suburbs, and the new patterns of urban living made possible by the automobile will continue to increase the number of cars in operation. Planning of the urban area must recognize that car ownership has become a necessary adjunct to home ownership, a key to widening opportunities for employment, a means of realizing recreational and other leisure time objectives, and an important factor in the loosening-up process that will gradually overcome outmoded patterns of congested living.

Public carriers, however, continue to be an essential part of the transportation system in large metropolitan areas. There is consequently a pressing need to revitalize transit and to maintain public transportation service where it can fulfill the specialized tasks it is most capable of performing. The principal task of transit will be to absorb home-to-work travel peaks. Public carriers will have to complement and supplement private transportation wherever the density of urban development and the concentration of urban travel dictate. Public transportation will continue to play an important role in the older central business districts, and along high-density, home-to-work travel routes close to the center. In these circumstances, the limited capacity of downtown highway and parking facilities in major cities will continue to make extensive use of the automobile impossible at the peak.

Improved bus transportation has been seen in Chapter IV to offer the greatest potential for public transportation from both an operation and economic standpoint, assuming a system of modern highways. The bus avoids the high cost of an exclusive right-of-way, and it can serve essentially the same transportation patterns established by the automobile. The inflexibility of rail operations, on the other hand, indicates that rail rapid transit and railroad commutation services will not be greatly expanded beyond facilities now in use. Modernization will be necessary, and important additions to transit will be required in the largest and oldest cities. A satisfactory transportation system for most urban areas, however, will have only a short mileage of rail routes. But these lines may play a key role in accommodating home-to-work peaks.

Automatic trains or other rapid transit innovations now promise to provide a high standard of transport service at reasonable cost. The application of these mass carrier techniques, as in San Francisco and Washington, will help to maintain high-density routes and high-density areas. The further expansion of motorized transport and the resulting changes in urban living and traveling, however, will predominate. Transport innovations will modernize all methods of movement, but their impact is more likely to be felt in road technology rather than rail, and, for the more distant future, in the air rather than on the ground. The helicopter, convertiplane, or other direct-lift aircraft will some day furnish the transportation service necessary to spread the urban traffic load over a wider area.

Highway and terminal requirements to meet the present and foreseeable future demands of motorized traffic, however, will continue to dominate the problem. Satisfactory transportation in urban areas calls for a complete system of expressways and the redesign of major surface streets to enable them to serve more effectively either for the movement of through traffic or for access to property. The attempt to serve both purposes at the same time has failed. It will be necessary to eliminate unneeded mileage and unnecessary intersections, to separate different types of traffic to minimize conflicts among incompatible street uses, and to develop new concepts of road design in conjunction with urban redevelopment and the planning of neighborhoods and commercial areas.

Only a total network of controlled-access expressways and parking facilities can provide a skeleton that will support the giant metropolis of the future. If only parts of the highway network are of satisfactory design, the skeleton is bound to collapse under the weight of the peak-hour movement attracted by expressway standards. It will also be necessary to combine the construction of main arteries of travel with adjacent land uses. Commercial enterprises serving the public, such as service stations, shopping centers, restaurants, motels, and terminals need to be located and designed as an integral part of the highway system. Chapter VII has pointed out that the acquisition of sufficient width of highway right-of-way can serve as the basis for planned commercial areas removed from the traveled way but readily accessible to it. Such an approach can combat the causes of suburban blight that have their roots in unplanned and unprotected roadsides. The need is for more effective legal and financial methods to make possible the reservation of highway rights-of-way that will be needed in the new suburbs, and the purchase of sufficient land in addition to what is needed specifically for the highway itself.

Revision of Financial Policy

A program to achieve better standards of transportation for urban areas would mean very substantial increases in transport facilities, including express highways, terminals, and transit systems. Yet it is clear from the discussion in Chapter V that the accomplishment of a self-supporting urban transportation system is well within the bounds of feasibility. Three steps are necessary. First, urban areas should be granted a fair share of state-collected highway user revenues, in the form of either cash grants from the states or by more adequate state construction programs in urban areas. Second, a more scientific pricing of transportation services is called for to maximize revenues and to achieve the most effective use of facilities. Third, the pooling of transportation revenues suggests a promising means of supporting high standards of service for the transportation system as a whole, including mass transit.

The low-fare policy that has led to low standards in the transportation field must be replaced by a transportation price policy that makes high standards possible. Depressed rates have preserved obsolete roads, antiquated railroads and transit equipment, and inadequate service. General tax support has been no solution. It has generally resulted in inadequate funds, uneconomical operations, and an absence of long-range physical and financial planning.

Thus far transit fare increases have been used mainly to cover rising costs rather than to provide better service, and reduced patronage has been the result. If fares were adequate to finance improved standards of service, however, the effect might be different. Express bus service at premium fares has actually promoted transit travel, and de luxe bus service has also indicated that higher fares can lead to increased traffic. More rate experimentation is needed, and more opportunity to experiment would be provided by eliminating control by the public utility commission over transit rates and placing this responsibility in a public transportation agency or authority.

Although use of the pricing mechanism to control transportation demand has not been attempted to any important degree, its potentials are promising. Pricing policies need to be established by one agency, however, if they are to be effective from the standpoint of either revenue production or traffic control. Tolls on urban expressways might be an effective method of minimizing less essential movement in the peak hour, and parking rates downtown might further assist in regulating traffic flow. For example, low parking rates for cars arriving downtown after the morning peak or for cars departing before the evening rush hour might be effective in reducing unnecessary concentrations of traffic. Joint parking and transit rates and pricing policies for public transportation based on mileage might be more remunerative and publicly acceptable. The elimination of free parking or parking meter charges at uneconomically low levels would add considerably to the support of the transportation system. This would also eliminate an important element of subsidy now encouraging the use of the automobile, and would provide a more equitable basis of competition between auto and transit. Transit fare increases designed to furnish better service might be more feasible if at the same time the automobile user were made to pay the full cost of driving.

Financing of urban transportation might be greatly assisted by a system approach in which all revenues were pooled and all operations together made to cover their costs. The use of highway revenues to help support improved peak-hour transit service might prove acceptable to highway users in the future. Automobiles and transit already share the cost of highways they use jointly. Transit operations contribute substantial sums to highway financing through the payment of motor vehicle tax revenues to the state. In many urban centers, transit operations also reduce the volume of new highway capacity required in the peak. From a total transportation viewpoint, the improvement of motor bus operations may mean a substantial saving in road expenditures or substantial relief for motorists attempting to make use of limited highway capacity. And logic would appear to favor using motor vehicle revenues generated by urban traffic to help support a metropolitan system of transportation rather than to subsidize little used rural roads.

A further reason for pooling transport revenues is suggested by the fact that transit provides an important standby service for motorists who make occasional use of public carriers when the family car is unavailable. As automobile ownership increases, more people may be expected to use the bus occasionally rather than regularly, and thus the standby role of the transit system will become increasingly important. If transit fares were increased to cover these standby costs, the burden would fall on automobile owners and non-owners alike. This would mean that non-owners, who generally represent groups least able to pay, would be incurring a daily surcharge to help support standby capacity for the sporadic rider.

The possibility of adapting toll financing to the urban area presents a specific method of pooling that might greatly strengthen the financial position of urban transportation. Tolls could provide quickly a complete system of expressways, transit, and terminals for the metropolitan area. Such a system might consist of a bond-financed network of expressways terminating in parking areas downtown and at fringe locations. Payment for use of the highway could be included in the parking fee rather than collected at the toll gate. Private vehicles in the close-in areas could be restricted to ramps leading to parking places only, while buses might be allowed to drive off the expressways and circulate on downtown surface streets. All costs of the system would be included in charges paid either on transit vehicles or at parking lots.

Managing the Transportation System

A major step toward planning and financing a satisfactory urban transportation system will be to establish the necessary governmental organization. Urban transportation must be removed from the administrative vacuum that has kept it from playing a full role in the development of better cities. Urban communities that divide the transportation problem into small parts cannot expect to get whole answers.

The need for organizing all available transportation facilities and for financing and operating them as a system is becoming increasingly apparent as we review the methods of supplying transport services today. Chapter VI has pointed out that the division of responsibility among political jurisdictions and among different agencies and departments within a single unit of government is the most obvious defect of current policy. The fact that roads and streets are provided by a number of jurisdictions in the metropolitan area is a frequent source of planning and financing difficulties, especially in the many instances where two or more states are involved. The large daily influx of commuters from outlying areas to the central city pose additional problems of supporting needed facilities on an equitable basis. Public transportation operations are often circumscribed by local government boundaries that impose economic burdens on both the carriers and the public. In many cases, political units have been made obsolete by the very transportation services they are attempting to furnish.

The administrative separation of different methods of transportation imposes a serious obstacle to effective community mobility. Decisions with respect to transit, for example, may have a controlling influence on the volume of automobile use, while pricing decisions governing parking may have a significant impact on transit patronage or highway requirements. The need for planning and financing all facilities with these interrelations in mind is apparent, but this is not possible with transportation responsibilities divided among separate agencies and jurisdictions.

The state highway departments have in many cases become an anachronism in the urban area. They are responsible for some of the most important travel routes in cities, and yet their limited jurisdiction precludes a broad approach to the needs of the city from the standpoint either of transportation or community planning. Urban highway work under the state is rapidly increasing in importance, but it is generally governed by the concept that city streets are merely connecting links in a state-wide system. It is little wonder that the city, confronted by the highly complex problem of accomplishing a total circulatory system properly related to over-all community goals has often found limited relief in state highway construction projects.

Transit is also the victim of partial remedies. The transit patron is not interested merely in the bus he rides but in the ride itself—which means the highway as well as the bus. Public transportation depends on both. As the transit company is not responsible for roads or their use—but only for the vehicle and its operation—the chances of getting good transit service are poor. Transportation policy needs to be aimed not simply at supplying the various elements of movement, but at improving standards of mobility.

A number of approaches toward integrating urban transportation functions have been noted in Chapter VI. No one of them has gone far enough to achieve, in any comprehensive way, the provision of good passenger service in the urban area. In the case of public transportation, the transit system has not been made part of a total transportation system, nor has any city solved the management and financial problems that the quest for reasonable standards of public transportation service introduce. It must be concluded that the attempt to overcome urban transportation problems with privately owned transit facilities is unrealistic. The provision of tolerable standards of service in the peak hours calls for much more equipment than can be profitably provided. At the same time, it is clear that neither public ownership of transit nor the transit authority has succeeded in providing satisfactory standards of service or financial strength. Deficits persist, and patronage continues downward on private and public lines alike. Only if transit were to be made part of a total transportation system, under unified policy direction or unified management, would the possibilities of more effective operations be realized.

It is equally clear that metropolitan areas are not organized to carry out an effective attack on their highway problems. Obviously, there must be greater local autonomy with respect to the location, design, and use of highways in urban areas to permit a closer relation between highway construction and urban development. To the extent that state highway departments continue to perform road construction work in cities, they must be better equipped to deal with the specialized problems encountered in urbanized areas. The establishment of strong urban divisions in state and federal highway bureaus has become increasingly important. Administrative machinery to achieve a co-ordinated transportation-community planning approach at the local level is also imperative if urban areas are to adapt to the automotive age.

The transportation difficulties of metropolitan areas, then, are not likely to be overcome by anything short of a complete alteration of administrative machinery. The need is for a physical plan and for investment and operating decisions designed to accomplish the plan. The solution might be a "transport authority," a "transportation district commission," or some form of metropolitan government. Some of the organizational experiments tried to date indicate what can be done. Facilities might be publicly owned and operated, publicly owned and privately operated, or privately owned and operated on a management contract basis. The essential requirement is that plans and policies should be uniform for the entire geographical area as well as for all relevant transportation facilities and services. This means uniform policies and integrated plans for major expressways and highways, related parking and terminal facilities, transit and railroad commuter services, taxi operations, and traffic engineering and control. It also means integrating the transportation function with metropolitan planning.

The Community Plan

The problem of achieving satisfactory standards of mobility for urban communities is only partly a transportation problem. The difficulties of urban mobility stem from more deep-seated causes, principally the concentration and haphazard development of urban communities. Chapter VII has furnished evidence of the need to do more than organize, plan, and finance additional transport capacity. It will be necessary to exert a positive control over the demand for transportation as well.

Two approaches have been indicated that can level the peaks of travel that are placing an impossible burden on the transportation system. One, designed to furnish immediate relief, is a community-wide program of staggered hours for working, shopping, and school. A spreading of the urban traffic load might prove highly advantageous to the city, the worker, and the economy as a whole. The cost of peak-hour highway and public transit capacity and the economic losses from traffic delays and personal annoyance are heavy. Actually, it may be costing more to accommodate the peak than it would cost if the length of the work day were reduced to promote the staggering of arrivals and departures.

The peak, in addition to being spread over more time, should be spread over more space. Cities can never solve their transportation problems if they continue to crowd too many people and too much economic activity into too little space. Congestion in the rush hours is inevitable as long as we insist on living in the suburbs, working downtown, and starting off at the same time to get to the same place. In these circumstances no transportation magic can make the journey to work a joy ride. We will have to avoid unmanageable transportation demand through the dispersal of population and economic activity, the preservation of open spaces, and the planning of land-use densities and arrangements.

Both population limits and geographical limits will have to be imposed on urban development if the metropolis is to avoid strangling in its own prosperity. There is increasing evidence of the need for directing more urban growth into new towns and existing smaller towns. This would seem preferable to the overcrowding that modern transportation now makes unnecessary, or to the endless sprawl that modern transport has made possible. But there is the further need for redeveloping existing urban centers to assure a generous balancing of developed land with open space, and for planning new suburban growth to assure the preservation of surrounding low-density land. Otherwise a solid build-up will ultimately deny easy access to the open country, and communities may become so unwieldy that the task of providing transportation and other community needs may destroy the advantages of urban living.

Redevelopment of existing urban communities and plans for new urbanization can help overcome peak-hour congestion by enabling people to live and work in the same areas, either close-in or on the periphery. Traffic can in this way be reduced by the elimination of unnecessary travel, a dispersal of the total volume of movement, and a reversal of peak-hour flows. Approaching the demand side of the problem offers a real hope of halting the endless race between traffic growth and the capacity of the transportation system. The fact that major development of the urban area has already taken place frequently discourages efforts that seem too late. Yet renewal of the city is constantly taking place, and comprehensive planning is giving cities a second chance.

The seeds of better community planning that are finding fertile ground in the decay of downtown, however, need to be transplanted to the fringes. Events that have left large areas of the central city in economic ruin are being re-enacted in the suburbs. Planless growth is adding to the transportation problems of the metropolitan area faster than the central city can hope to overcome its past mistakes through redevelopment. Whether in the city or in the suburbs, the isolation of urban planning from transportation planning has proved impractical. Both problems are compounded by the attempted separation. Satisfactory transportation is impossible without comprehensive planning that exploits to the fullest the relations between good transportation and good communities.

The federal government has promoted metropolitan area planning through recent housing legislation requiring urban renewal projects to be related to a comprehensive urban plan. Planning of transportation facilities has also been furthered through transportation surveys jointly supported by federal, state, and local governments, with federal funds supplied through federal-aid highway legislation. A total approach is obviously desirable, and more effective federal encouragement of metropolitan area planning is badly needed, especially in connection with the accelerated program of urban highway development sponsored by the national government. Road construction can have highly damaging effects in urban areas unless there are area-wide community development plans to guide it. Approval of federal projects has now been made contingent on adherence to an over-all transportation plan. There is still need to assure a satisfactory relation to comprehensive community planning.

An effective solution to the urban transportation problem, then, should meet three tests. First, it should be functionally comprehensive by including all forms of transportation applicable to the problem. Second, it should be comprehensive geographically by including not only the city but the metropolitan area and all the affected region. Third, it should be comprehensive from a planning standpoint by assuring that transportation is used to promote community goals, and that community plans make satisfactory transportation possible.

This latter test is the most important. The basic need is to achieve satisfactory conditions of living. In the cities that made American industry the most prosperous in the world, slums and blight are an anomaly that needs to be attacked with all the resources at hand. Transportation development that merely helps to move us more expeditiously through areas of urban decay misses the mark.

The transportation industries, operating under archaic public policies, have failed to contribute their full potential to the building of better communities. Yet an attack on transportation inadequacies, broadly viewed, is not something apart from an attack on the inadequacies of the city. Transportation facilities that now provide the escape from undesirable urban conditions can help to overcome these conditions. Decisions governing transportation can exert an overriding influence on future patterns of urban development. Transportation facilities can blight the area through which they pass, or they can restore it. They can further the development of park and recreation lands and other objectives of the city, or they can simply carry traffic. And in doing so, they can support pleasant neighborhoods and prosperous communities, or they can nullify efforts to attain a higher standard of urban living.

In American communities most of the housing and commercial developments now taking place make little sense in relation to the ways that people move today or will be moving in the future. At one extreme, we are preserving the old congested way of urban living as if the technological innovations in transportation and communications did not exist. At the other extreme, we are reacting against high densities by substituting endless sprawl, with no effort to control growth, preserve open spaces, or apply our new-found mobility to the enhancement of urban life.

In a nation that is both motorized and urbanized, there will have to be a closer relation between transportation and urban development. We will have to use transportation resources to achieve better communities and community planning techniques to achieve better transportation. The combination could launch a revolutionary attack on urban congestion that is long overdue.

A Note on Lessons for Developing Countries

Rapid industrialization and urbanization are creating in the big cities of newly developing countries the same kinds of transportation problems faced by the metropolitan areas of the United States. Congestion in major urban centers is not only reducing the efficiency of economic activities but is creating a heavy backlog of capital investments for transport and other social overheads. These requirements are a serious burden in countries where capital is scarce.

Asia has more large cities and more people living in them than either North America or Europe. In South America a major part of the urban population is concentrated in a small number of great metropolitan areas. Much of this urbanization is not the product of industrialization, but has occurred without it, and has created extensive slums and other conditions posing a threat to development.[1]

The urban problems of developing countries are rapidly increasing in severity. Programs of industrialization, rapid growth of population, and the accelerating rate of migration from farm to city are all adding to the concentration of activity in urban areas. As a result, in many underdeveloped countries the rate of urban growth is far greater than in the more developed areas of the world. Between 1900 and 1950, for example, while the population of large cities in North America and Europe increased 160 per cent, the urban population of Asia grew 444 per cent. Many less developed areas have a low ratio of urban to total population, however. Only 8 per cent of Asia's people are in cities of 100,000 or more, compared with 21 per cent in Europe and 29 per cent in North America. There

[1] Philip H. Hauser, "Implications of Population Trends for Regional and Urban Planning in Asia." ECAFE Working Paper No. 2. Tokyo, July–August 1958, p. 9.

is still time, then, to adopt measures that will reduce transport problems created by unrestrained and planless urban growth. Help in doing this can be provided by the lessons learned from Western experience. Past successes and mistakes indicate what policies should be avoided and what new solutions are available to countries in earlier stages of development.

The nature of the problem is illustrated by the situation in India. If present trends continue, a third of India's population will be in urban areas by the year 2000. The largest metropolis, Calcutta, will have between 12 and 16 million inhabitants in 1970.[2] One reason for this heavy concentration is that there are too few large cities in India, and consequently each one has a vast rural hinterland from which the rural unemployed are migrating. Calcutta is the only major metropolis in the entire area of West Bengal, Bihar, Orissa, and Assam.[3]

The great size of both Calcutta and Bombay reflects the fact that the major part of India's ocean trade passes through these port cities. In addition, new industries are constantly being attracted by the extensive local market, by the ready supply of labor, and by the existing infrastructure of urban facilities and services. If the undesirable trends toward further congestion are to be avoided, new ports must be built, industrial expansion will have to be promoted elsewhere, and certain types of growth in the largest cities will have to be curtailed through government taxation and investment policy.

Modern transport and communications provide one of the means by which a more dispersed pattern of urbanization can be attained. The telephone, the airplane, road

[2] Kingsley Davis, "Urbanization in India: Past and Future," in *India's Urban Future,* edited by Roy Turner, University of California Press, 1962.

[3] "Industrial Location and Urbanization with Special Reference to Calcutta" (Annex 5), *Country Assistance Program: India,* Agency for International Development, Department of State, August 1964, pp. 5/1–5/12.

transport, and power distribution are all instruments of dispersal. They can help to overcome the centering influence of major ports and railways. But their use for this purpose will have to be combined with development plans for more moderate-sized communities. This means the planned location of new industries, improved public services, better educational facilities, new housing construction, the establishment of agricultural processing factories, and a variety of amenities to enhance the environment of smaller cities.

The planned development of new or expanded communities of moderate size makes it possible to attack the transport problem effectively from the start. The necessary transport routes and terminals can be set aside to meet projected traffic needs, and land uses can be planned to avoid over-concentration, to preserve open space and recreation areas, and to relate employment opportunities and housing in ways that will minimize home-to-work travel. Of key importance is the reserving of land for anticipated public purposes at an early stage.

The fact remains, however, that it will be quite impossible to avoid further increments of growth in those places that are already overgrown. The task of coping with traffic congestion in these areas will continue to intensify, especially since further economic growth is bound to increase reliance on motorized transport. The situation calls for a master plan for urban growth, for each urban area, and for the transportation systems that will be needed in the long run to provide for the movement of people and goods.[4]

The effort to provide capacity to move will be defeated unless there is adherence to planned densities and arrangements of the land uses that generate traffic. The type and density of allowable development should be based on the capacities to be made available for passenger and freight movement. Open space should be preserved not only for recreation, but to compensate for the traffic generation of built-up land. It may also be advantageous to preserve a

4 Wilfred Owen, *Strategy for Mobility*, 1964, pp. 79–82.

belt of agricultural land around the big city to set limits to the urbanized area and at the same time to make possible the production of food for local consumption.

With respect to transport planning itself, suitable locations will be required for main routes, markets, and terminals, to keep freight traffic out of the urban center. Some areas in the older cities will have to be reserved primarily for pedestrians, while animal-drawn vehicles will have to be barred from major traffic arteries. Routes will need to be designated for the exclusive use of buses, and some cities will require special roadways for bicycles. The construction of bypass routes and the use of controlled-access highways will permit important long-run solutions.

The magnitude of the transport problem and the outlook for future years are indicated in a recent study of Bombay.[5] During the 1950's the number of motor vehicles in that city more than doubled. At the same time, mass transportation patrons were increasing at an average annual rate of 4 per cent. In 1963 a total of 1.8 million passengers were being carried daily by bus on 600 miles of city routes, while suburban railways were carrying another 1.6 million passengers daily. Needed express highways are being built at heavy cost, but this is only a small part of the investment needed. Other requirements include new public transit facilities, freight terminals, warehousing and storage facilities, and measures to handle increasing volumes of automotive traffic and vehicle parking.

But the details of urban planning and transport systems planning should not obscure the need for planning the location, size, and patterns of urban settlement on a national scale in relation to national economic development plans. It should be possible, through central government planning of transport, industry, and public works, and through local planning of the urban area, to guide the growth of cities. Removal of the capital of Brazil from congested Rio de Janeiro to the new city of Brasília, and

[5] Wilbur Smith and Associates, *Bombay Traffic and Transportation Study*, 1963, Vol. 1.

the shift of Pakistan's capital from Karachi to Islamabad, illustrate the extreme steps being taken to avoid over-concentration and to seek a balanced regional development. But ideally the solution to excessive urban concentration is not to design one spectacular remedy for one unmanageable situation. The task is rather to plan the urbanization process on a continuing basis.

One can be sure from current growth trends that urban requirements will give rise to the major demands for transport investment in developing nations in the years immediately ahead. Already there is a shift in emphasis to the growing needs of the urban economy. In carrying out the necessary urban development effort it will be advances in technology, and especially in transport and communications, that will make possible fresh solutions to metropolitan transportation problems. A closer examination of the transportation difficulties that seem so intractable in Chicago and New York may help to mitigate the problems of Bangkok and Beirut, and of new cities in the developing world yet to feel the effects of the urban-industrial revolution.

Appendix

TABLE 1. Population of the United States by Type of Residence, 1950 and 1960

Area	1950	1960	Increase, 1950–60, in Number	Per Cent
Total United States	151,326	179,323	27,997	18.5
Standard Metropolitan Areas	89,317	112,885	23,568	26.4
Central cities	52,386	58,004	5,618	10.7
Outside central cities	36,931	54,881	17,950	48.6
Other Territory	62,009	66,438	4,429	7.1

Source: U. S. Bureau of the Census, *Census of Population: 1960, Number of Inhabitants,* pp. XXVI and 1–3.

TABLE 2. Metropolitan Areas of the United States, 1960

Population Class	Number of Metropolitan Areas	Millions of People
3,000,000 and over	5	31.8
1,000,000–2,999,000	19	29.8
500,000– 999,000	29	19.2
250,000– 499,000	48	15.8
100,000– 249,000	89	14.5
under 100,000	22	1.8
Total	212	112.9

Source: U. S. Bureau of the Census, "Number of Inhabitants," *U. S. Census of Population, 1960: United States Summary,* Table 0, p. xxv.

TABLE 3. Population in the Twenty-Five Largest U. S. Metropolitan Areas, 1960

Metropolitan Area	Population		Population Density	
	Metropolitan Area	Central City	Metropolitan Area	Central City
New York–Northeastern N.J.[a]	14,114,927	8,743,015	7,462	23,321
Los Angeles–Long Beach, Calif.	6,488,791	2,823,183	4,736	5,638
Chicago–Northwestern Ind.	5,959,213	3,898,091	6,209	12,959
Philadelphia, Pa.–N.J.	3,635,228	2,002,512	6,092	15,743
Detroit, Mich.	3,537,709	1,670,144	4,834	11,964
San Francisco–Oakland, Calif.	2,430,663	1,107,864	4,253	11,013
Boston, Mass.	2,413,236	697,197	4,679	14,586
Washington, D.C.	1,808,423	763,956	5,308	12,442
Pittsburgh, Pa.	1,804,400	604,332	3,437	11,171
Cleveland, Ohio	1,784,991	876,050	3,042	10,789
St. Louis, Mo.–Ill.	1,667,693	750,026	5,160	12,296
Baltimore, Md.	1,418,948	939,024	6,441	11,886
Minneapolis–St. Paul, Minn.	1,377,143	796,283	2,095	7,326
Milwaukee, Wis.	1,149,997	741,324	2,934	8,137
Houston, Texas	1,139,678	938,219	2,647	2,860
Buffalo, N.Y.	1,054,370	532,759	6,582	13,522
Cincinnati, Ohio–Ky.	993,568	502,550	4,101	6,501
Dallas, Texas	932,349	679,684	1,441	2,428
Kansas City, Mo.–Kans.	921,121	475,539	3,262	3,664
Seattle, Wash.	864,109	557,087	3,626	6,295
Miami, Fla.	852,705	291,668	4,657	8,529
San Diego, Calif.	836,175	573,224	3,033	2,979
Denver, Colo.	803,624	493,887	4,824	6,956
Atlanta, Ga.	768,125	487,455	3,125	3,802
Providence, R.I.–Mass.	659,542	288,499	3,508	10,887

Source: Derived from Table 22, United States Census, 1960, United States Summary, Number of Inhabitants.

[a] Population—New York City: 7.8 million; Manhattan: 1.7 million; and population density—New York City: 24,697; Manhattan:

TABLE 4. Resident and Daytime Population of Major Cities, 1950[a]
(In thousands)

City	Resident Population	Daytime Population	Percentage Increase of Day over Resident Population
Baltimore, Md.	950	1,071	13
Boston, Mass.	801	1,075	34
Chicago, Ill.	3,621	4,252	17
Cincinnati, Ohio	504	621	23
Cleveland, Ohio	915	1,086	19
Detroit, Mich.	1,850	2,182	18
Jersey City, N.J.	299	296	−1
Los Angeles, Calif.	1,970	2,591	31
Milwaukee, Wis.	637	731	15
Newark, N.J.	439	885	102
New York, N.Y.	7,892	8,892	13
Pittsburgh, Pa.	677	1,012	49
San Francisco, Calif.	775	1,012	31
Washington, D.C.	802	1,009	26
Philadelphia, Pa.	2,072	2,465	19

[a] *1950 Census of Population,* and Civil Defense Administration, *Population of 89 Major Cities of the Critical Target Areas* (1953).

TABLE 5. Percentage Distribution of the Resident Population of the City of Chicago, by Distance Zone from the Center of the City[a]

Year	0–2 mi.	2–4 mi.	4–6 mi.	6–8 mi.	8 mi. or more	All Zones
1900	21.5	43.2	16.6	12.9	5.8	100.0
1910	16.5	32.6	28.4	13.3	9.2	100.0
1920	10.2	26.5	31.5	18.5	13.3	100.0
1930	6.6	19.6	28.4	24.4	21.0	100.0
1940	5.9	18.6	28.5	25.1	21.9	100.0
1950	6.1	18.5	27.2	23.9	24.3	100.0

[a] *Growth and Redistribution of the Resident Population in the Chicago Standard Metropolitan Area,* A Report by the Chicago Community Inventory to the Office of the Housing and Redevelopment Coordinator and the Chicago Plan Commission (1954), pp. 18–19.

TABLE 6. Population Changes in Chicago, 1930 to 1950[a]

Miles from Center of City	Percentage Change in Population	
	1930–40	1940–50
Within 1 mile	− 7.2	18.0
1–2 miles	−10.7	9.9
2–3 miles	− 8.4	5.7
3–4 miles	− 1.5	6.5
4–5 miles	0.6	1.4
6–8 miles	3.9	1.4
8–10 miles	4.9	10.1
10 or more	4.7	33.1
City of Chicago	0.6	6.6

[a] *Growth and Redistribution of the Resident Population in the Chicago Standard Metropolitan Area*, A Report by the Chicago Community Inventory, to the Office of the Housing and Redevelopment Coordinator and the Chicago Plan Commission (1954), pp. 18–19.

TABLE 7. Percentage Distribution of the Population of Philadelphia by Distance from the Center of the City, 1900–50[a]

Miles from Center of City	1900	1920	1940	1950
0– 1	9.3	5.5	3.2	3.1
0– 2	24.0	17.8	12.3	11.0
0– 5	72.6	68.9	55.3	50.1
5–10	11.2	15.5	26.6	29.9
10–18	9.9	10.4	12.4	13.9
18–25	6.3	5.2	5.7	6.2
0–25	100.0	100.0	100.0	100.0

[a] Data from Hans Blumenfeld, "The Tidal Wave of Metropolitan Expansion," *Journal of the American Institute of Planners* (Winter 1954), p. 13.

TABLE 8. Means of Travel to and from Place of Work, Selected U. S. Cities[a]

City	Private Automobile or Car Pool[b]		Public Transportation[b]	
	Central City	Suburb	Central City	Suburb
New York, N.Y.	20	63	65	23
Los Angeles–Long Beach, Calif.	75	86	13	4
Chicago, Ill.	45	70	43	16
Philadelphia, Pa.	42	70	44	13
Detroit, Mich.	68	86	23	5
San Francisco, Calif.	51	79	33	8
Boston, Mass.	38	67	43	18
Pittsburgh, Pa.	49	69	34	15
Wash. D.C.–Md.–Va.	43	77	42	12
Cleveland, Ohio	58	77	32	14
Baltimore, Md.	56	78	30	7
Newark, N.J.	46	68	42	20
Minneapolis–St. Paul, Minn.	64	83	21	5
Seattle, Wash.	68	85	19	3
Cincinnati, Ohio	63	76	24	12
San Diego, Calif.	64	73	8	2
Denver, Colo.	71	84	15	3
Tampa–St. Petersburg, Fla.	77	81	10	2
Phoenix, Ariz.	85	76	5	3
San Jose, Calif.	82	82	7	4
Fort Worth, Tex.	80	88	10	1
Akron, Ohio	76	86	13	4
Oklahoma City, Okla.	83	85	6	1
Sacramento, Calif.	75	87	11	2
Honolulu, Hawaii	71	58	16	2
Flint, Mich.	81	89	7	1
Fresno, Calif.	84	76	5	1
Wichita, Kans.	84	88	6	1
Standard Metropolitan Areas (Number of Population)				
Over 1,000,000	48	76	39	12
500,000 to 1,000,000	68	78	18	6
300,000 to 500,000	70	80	15	4
250,000 to 300,000	72	76	13	4
200,000 to 250,000	72	75	11	3
Under 100,000	76	64	6	3

Source: Data supplied by the Bureau of Public Roads, "Selected Statistics by Standard Metropolitan Statistical Areas for Use in Transport Planning," August 1964, Tables 6, 6A, 6B, & 6C (compiled by Edmond L. Kanwit).

[a] Survey conducted during census week, 1960.

[b] The balance other than private automobile, car pool, and public transportation represents "walked to work, worked at home, and others."

TABLE 9. Transportation Expenditures by Urban
Consumers, Selected Cities, 1950[a]

City	Transport Expenditures as Percentage of Total Consumer Expenditures	Automobile Transportation		All Other Transportation	
		Amount	Per Cent	Amount	Per Cent
Baltimore, Md.	12.7	$403	10.3	$ 95	2.4
Boston, Mass.	9.9	323	7.5	103	2.4
Chicago, Ill.	12.7	497	10.1	129	2.6
Cleveland, Ohio	14.3	560	12.0	108	2.3
Los Angeles, Calif.	16.4	692	14.8	74	1.6
New York, N.Y.	8.5	294	6.0	121	2.5
Northern New Jersey Area	11.2	459	9.7	72	1.5
Philadelphia–Camden	10.5	353	8.1	105	2.4
Pittsburgh, Pa.	13.3	499	11.1	98	2.2
San Francisco–Oakland, Calif.	14.1	548	12.2	86	1.9
St. Louis, Mo.	14.4	523	12.3	88	2.1
Atlanta, Ga.	14.1	463	12.3	67	1.8
Birmingham, Ala.	14.8	421	12.9	61	1.9
Cincinnati, Ohio	14.2	515	12.3	79	1.9
Indianapolis, Ind.	14.7	504	13.1	61	1.6
Kansas City, Mo.	14.7	523	13.1	65	1.6
Louisville, Ky.	14.1	465	12.4	65	1.7
Miami, Fla.	14.4	576	12.5	87	1.9
Milwaukee, Wis.	13.9	509	11.8	90	2.1
Minneapolis–St. Paul, Minn.	15.0	587	13.3	75	1.7
New Orleans, La.	12.7	328	9.8	98	2.9
Norfolk–Portsmouth, Va.	14.1	437	12.0	77	2.1
Omaha, Neb.	14.0	473	11.9	84	2.1
Portland, Ore.	16.8	620	15.0	75	1.8
Providence, R.I.	10.7	358	9.2	60	1.5
Seattle, Wash.	16.0	622	13.7	103	2.3

[a] U. S. Bureau of Labor Statistics, *Family Income, Expenditures and Savings in 1950,* Bulletin 1097 (June 1953), pp. 17–41.

TABLE 10. Index of Transportation Trends in Relation
to Economic Growth 1940–63
(1940 = 100)

Year	Gross National Product[a]	Total Population	Urban Population[b]	Industrial Employment	Consumer Transport Expenditures[a]	Automobile Registrations	Transit Riders
1940	100.0	100.0	100.0	100.0	100.0	100.0	100.0
1941	115.7	101.0	100.9	108.6	110.4	107.9	107.6
1942	129.7	102.1	101.2	117.2	64.6	101.8	138.3
1943	144.2	103.5	101.7	119.5	59.0	94.7	170.9
1944	154.5	104.7	100.3	118.8	59.0	93.0	178.9
1945	152.6	105.9	100.1	116.5	66.7	93.9	181.1
1946	137.3	107.0	111.3	123.6	109.7	102.7	182.3
1947	137.2	109.1	112.8	131.0	126.4	112.2	174.3
1948	142.4	111.0	113.8	134.4	136.1	121.3	164.9
1949	142.2	112.9	116.4	133.4	156.9	132.7	145.3
1950	154.6	114.8	120.6	138.1	179.2	146.8	131.9
1951	166.1	116.8	123.6	142.1	165.3	155.4	122.7
1952	171.8	118.8	126.6	143.5	166.0	159.5	114.6
1953	179.3	120.8	129.6	146.5	188.2	169.0	105.1
1954	176.4	122.9	132.7	144.1	186.8	176.4	93.9
1955	190.8	125.1	135.8	148.7	—[c]	189.8	87.6
1956	194.8	127.3	138.9	153.1	231.3	197.3	83.5
1957	198.5	129.7	142.0	154.8	241.0	203.5	79.5
1958	195.0	131.9	145.1	153.0	217.4	206.9	74.2
1959	208.3	134.1	148.2	157.3	250.7	216.7	72.9
1960	213.8	136.8	151.3	160.5	259.0	224.0	71.4
1961	217.6	139.1	154.4	161.5	247.9	230.2	68.8
1962	231.5	141.3	157.5	165.2	273.6	239.8	67.7
1963	239.4	143.3	160.6	168.1	288.9	250.1	65.7

Sources: *Economic Report of the President* (January 1965), pp. 192, 213, 214; Automobile Manufacturers Association, *Automobile Facts and Figures,* 1964, p. 18; U. S. Department of Commerce, *National Income. A Supplement to the Survey of Current Business,* 1954, pp. 206–7; *Survey of Current Business,* July 1955, p. 19; *Survey of Current Business,* July 1963, p. 16; number of transit riders from the American Transit Association.

[a] In 1954 dollars.
[b] Urban population estimates for the years between the decennial censuses are derived by adding one-tenth of the annual rate of population growth between 1940–50 and 1950–60 to the base.
[c] Not available.

238   METROPOLITAN TRANSPORTATION PROBLEM

TABLE 11. Fatality Rates on Highways with Controlled
and Uncontrolled Roadsides[a]

State	Highway	Period	Fatality Rate Per 100 Million Vehicle Miles
Maine	U. S. Route 1	1948	22.3
	Maine Turnpike	1948–50	2.8
Connecticut	Boston Post Road	1940–48	10.7
	Merritt Parkway	1940–48	3.4
New Jersey	3- and 4-lane undivided highways	1935–48	14.0
	Route 4, divided highway with grade separation	1935–48	4.0
	New Jersey Turnpike	1952	5.8
Virginia	Shirley Highway	1952	4.4
	U. S. Route 1	1952	17.9
California	Rural state highways	1947–51	9.0
	Arroyo-Seco Parkway	1947–51	1.5
	Hollywood Freeway	1948–51	2.5

[a] John W. Gibbons, "Economic Costs of Traffic Congestion," *Urban Traffic Congestion,* Highway Research Board Bulletin 86 (1954), p. 17.

TABLE 12. Status of Interstate Highway System in
1964 in Relation to Urban Character of States

State	Urban Population as Per Cent of Total	Per Cent of Interstate System Completed
New Jersey	88.6	39.5
Rhode Island	86.4	36.4
California	86.4	41.6
New York	85.4	70.9
Massachusetts	83.6	63.1
Illinois	80.7	48.9
Connecticut	78.3	67.3
Hawaii	76.5	—[a]
Texas	75.0	47.1
Utah	74.9	16.2
Arizona	74.5	53.3
Florida	73.9	37.4
Colorado	73.7	43.0

State	Urban Population as Per Cent of Total	Per Cent of Interstate System Completed
Ohio	73.4	58.8
Michigan	73.4	73.0
Maryland	72.7	67.1
Pennsylvania	71.6	51.9
Nevada	70.4	43.3
Washington	68.1	46.3
Missouri	66.6	52.7
New Mexico	65.9	41.7
Delaware	65.6	41.6
Wisconsin	63.8	67.2
Louisiana	63.3	22.8
Oklahoma	62.9	58.9
Indiana	62.4	44.8
Oregon	62.2	79.3
Minnesota	62.2	26.7
Kansas	61.0	66.3
New Hampshire	58.3	62.0
Wyoming	56.8	48.2
Virginia	55.6	38.4
Georgia	55.3	36.8
Alabama	54.8	34.4
Nebraska	54.3	43.1
Iowa	53.0	46.8
Tennessee	52.3	30.3
Maine	51.3	53.6
Montana	50.2	32.7
Idaho	47.5	45.2
Kentucky	44.5	37.0
Arkansas	42.8	27.4
South Carolina	41.2	50.1
North Carolina	39.5	48.7
South Dakota	39.3	45.9
Vermont	38.5	27.1
West Virginia	38.2	30.8
Alaska	37.9	—
Mississippi	37.7	37.7
North Dakota	35.2	55.3

Source: The National System of Interstate and Defense Highways, Bureau of Public Roads, U. S. Department of Commerce, Table I.

[a] 69.8 per cent of the total designated system mileage under way as of December 1964.

TABLE 13. Trends in Financing City Streets, 1940-63 (In millions)

Year	Local Expenditures: From Local Taxes	Local Expenditures: From State-aid[a]	State-Federal Expenditures[b]	State-Federal Expenditures and State-aid	Total Expenditures on Urban Streets	Total Urban Expenditures as Percentage of Total Streets and Roads	Total State-Federal Funds Spent on City Streets as Percentage of Total Highway-User Revenue	Total State-Federal Funds Spent on City Streets as Percentage of Total Urban Expenditures
1940	$ 331	$ 71	$ 209 a	$ 280	$ 611	23.7	24.7	45.8
1941	320	57	145 c	202	522	24.2	16.1	38.7
1942	274	58	88 c	146	420	23.7	12.9	34.8
1943	231	52	31	83	314	23.4	8.5	26.4
1944	255	50	27	77	332	24.6	7.5	23.2
1945	257	52	28	80	337	23.6	6.9	23.7
1946	308	86	41	127	435	21.3	8.2	29.2
1947	457	114	118	232	689	24.0	13.6	33.7
1948	564	120	181	301	865	25.1	15.6	34.8
1949	634	165	284	449	1,083	27.7	20.2	41.5
1950	633	175	336	511	1,144	27.5	21.0	44.7
1951	667	179	372	551	1,218	26.7	20.6	45.2
1952	746	196	402	598	1,344	26.6	19.8	44.5
1953	788	226	517	743	1,531	27.2	22.7	48.5
1954	834	259	654	913	1,747	26.6	25.8	52.3
1955	895	274	751	1,025	1,920	27.7	26.0	53.4
1956	1,006	290	989	1,279	2,285	28.8	29.7	56.0
1957	1,085	334	1,228	1,562	2,647	30.1	34.5	59.0
1958	1,166	392	1,334	1,726	2,892	29.5	36.7	59.7
1959	1,163	418	1,560	1,978	3,141	30.6	39.3	63.0
1960	1,210	474	1,476	1,950	3,160	31.1	36.2	61.7
1961	1,308	459	1,729	2,188	3,496	32.3	39.6	62.6
1962	1,362	496	2,101	2,597	3,959	34.1	44.4	65.6
1963	1,401	503	2,190	2,693	4,094	33.2	44.3	65.8
1964 d	1,458	525	2,448	2,973	4,431	32.4	45.6	67.1

Source: Highway Finances, 1921–1962, Highway Statistics Division, Bureau of Public Roads, Department of Commerce, Table HF-1 and HF-2, issued December 1964.

a State grants-in-aid and direct expenditures on local municipal streets by state.
b Includes direct expenditures on municipal extensions of state highways only.
c Includes WPA expenditures.
d Preliminary.

TABLE 14. Percentage of Passengers Entering Central
Business District by Mass Transportation on Weekdays

Selected Cities	Percentage Travel by Mass Transit	Date of Survey
Chicago, Ill.[a]	85	1960
New York, N.Y.[a]	85	1962
Newark, N.J.	80	1960
Philadelphia, Pa.	64	1953
Richmond, Va.	62	1955
Atlanta, Ga.	60	1953
Boston, Mass.	58	1954
San Antonio, Texas	57	1955
Dayton, Ohio	56	—[b]
Cleveland, Ohio	56	1960
Youngstown, Ohio	54	1950
Pittsburgh, Pa.	52	1953
Detroit, Mich.	47	1952
New Orleans, La.	47	1953
San Francisco, Calif.	46	1959
Milwaukee, Wis.	45	1955
Minneapolis, Minn.	42	1955
Dallas, Texas	38	—[b]
Providence, R.I.	37	1954
Wilmington, Del.	36	1947
Seattle, Wash.	35	1954
Los Angeles, Calif.	34	1960
Cincinnati, Ohio	33	1951
Washington, D.C.	32	1953
Spokane, Wash.	30	—[b]
Columbus, Ohio	30	1955
Kansas City, Mo.	28	1954
Louisville, Ky.	26	1953
Salt Lake City, Utah	25	—[b]
Springfield, Ill.	23	1948
Houston, Texas	23	1963

Source: Data supplied by the American Transit Association. The central
business district data do not reflect the fact that in outlying areas and on
weekends major dependence is on automobile.

[a] These figures are based on the total number of all transit passengers
entering the central business district and not exclusively mass transit.
[b] Not given.

TABLE 15. Consumer Transportation Expenditures
(In millions of 1954 dollars)

Year	Total Expenditures	Private Automobile	Local Public Carrier	Intercity Public Carrier
1929	12,357	9,675	1,813	869
1930	10,419	7,902	1,785	732
1931	9,511	7,124	1,751	637
1932	8,561	6,312	1,690	559
1933	8,900	6,775	1,607	518
1934	9,655	7,529	1,599	527
1935	10,866	8,677	1,626	564
1936	12,487	10,094	1,721	672
1937	12,804	10,389	1,711	703
1938	11,311	8,942	1,691	679
1939	12,937	10,423	1,785	730
1940	14,372	11,815	1,825	732
1941	15,891	13,241	1,842	808
1942	9,292	6,018	2,175	1,099
1943	8,522	4,398	2,532	1,591
1944	8,525	4,439	2,516	1,570
1945	9,641	5,623	2,459	1,559
1946	15,772	11,770	2,550	1,452
1947	18,191	14,558	2,372	1,261
1948	19,581	15,992	2,363	1,226
1949	22,608	19,097	2,352	1,159
1950	25,834	22,449	2,308	1,077
1951	23,779	20,402	2,217	1,160
1952	23,869	20,463	2,214	1,192
1953	27,092	23,698	2,193	1,201
1954	26,928	23,682	2,150	1,096
1955ᵃ	—	—	—	—
1956	33,288	30,144	1,946	1,198
1957	34,705	31,594	1,885	1,226
1958	31,281	28,298	1,795	1,188
1959	36,089	32,990	1,818	1,281
1960	37,345	34,170	1,817	1,358
1961	35,690	32,515	1,768	1,407
1962	39,449	36,145	1,783	1,521
1963	41,638	38,346	1,758	1,534

Source: U. S. Department of Commerce, *National Income,* A supplement to the *Survey of Current Business* (1954), pp. 206–7 for 1929–53 data; U. S. Department of Commerce, *Survey of Current Business,* July 1955, p. 19 for 1954 data; and *Survey of Current Business,* July 1963, p. 16.

ᵃ Not available.

TABLE 16. Trends in Methods of Transit, 1905–63
(In billions of total passengers)

Year	Street Car	Rapid Transit	Trolley Coaches	Motor Buses	Total
1905	5.0	–	–	–	5.0
1907	8.9	0.7	–	–	9.5
1912	11.1	1.0	–	–	12.1
1917	13.2	1.3	–	–	14.5
1918	12.9	1.4	–	–	14.2
1919	13.4	1.5	–	–	14.9
1920	13.7	1.8	–	–	15.5
1921	12.7	1.9	–	–	14.6
1922	13.4	1.9	–	0.4	15.7
1923	13.6	2.1	–	0.7	16.3
1924	13.1	2.2	–	1.0	16.3
1925	12.9	2.3	–	1.5	16.7
1926	12.9	2.4	–	2.0	17.2
1927	12.5	2.5	–	2.3	17.2
1928	12.0	2.5	0.004	2.5	17.0
1929	11.8	2.6	0.005	2.6	17.0
1930	10.5	2.6	0.02	2.5	15.6
1931	9.2	2.4	0.03	2.3	13.9
1932	7.6	2.2	0.04	2.1	12.0
1933	7.1	2.1	0.05	2.1	11.3
1934	7.4	2.2	0.07	2.4	12.0
1935	7.3	2.2	0.1	2.6	12.2
1936	7.5	2.3	0.1	3.2	13.1
1937	7.2	2.3	0.3	3.5	13.2
1938	6.5	2.2	0.4	3.5	12.6
1939	6.2	2.4	0.4	3.9	12.8
1940	5.9	2.4	0.5	4.2	13.1
1941	6.1	2.4	0.7	4.9	14.1
1942	7.3	2.6	0.9	7.2	18.0
1943	9.2	2.7	1.2	9.0	22.0
1944	9.5	2.6	1.2	9.6	23.0
1945	9.4	2.7	1.2	9.9	23.3
1946	9.0	2.8	1.3	10.2	23.4
1947	8.1	2.8	1.4	10.3	22.5
1948	6.5	2.6	1.5	10.7	21.4
1949	4.9	2.3	1.7	10.2	19.0
1950	3.9	2.3	1.7	9.4	17.2
1951	3.1	2.2	1.6	9.2	16.1
1952	2.5	2.1	1.6	8.9	15.1
1953	2.0	2.0	1.6	8.2	13.9
1954	1.5	1.9	1.4	7.6	12.4
1955	1.2	1.9	1.2	7.3	11.6
1956	0.9	1.9	1.2	7.1	11.0
1957	0.7	1.8	1.0	6.9	10.4
1958	0.6	1.8	0.8	6.5	9.8

Year	Street Car	Rapid Transit	Trolley Coaches	Motor Buses	Total
1959	0.5	1.8	0.8	6.5	9.6
1960	0.5	1.9	0.7	6.4	9.4
1961	0.4	1.9	0.6	6.0	8.9
1962	0.4	1.9	0.6	5.9	8.7
1963	0.3	1.8	0.4	5.8	8.4

Source: Data supplied by the American Transit Association.

TABLE 17. Person Trips Leaving Philadelphia Central Business District, Spring Weekday, 1955[a]

Mode of Transportation	Peak-Hour (5–6 p.m.)			Total Day (24 hours)	
	Trips	Percentage of Peak Total by all Modes	Percentage of Daily Total by all Modes	Trips	Percentage of Daily Total by all Modes
Transit and Rail	126,592	71.5	23.6	537,700	50.0
PTC surface	35,556	20.1	16.5	216,000	20.1
Rapid transit	62,589	35.3	25.5	245,700	22.8
Suburban rail	21,912	12.4	44.5	49,200	4.6
Interstate bus	6,535	3.7	24.4	26,800	2.5
Auto, Taxi, Truck	45,066	25.5	9.4	480,300	44.6
Auto and taxi	42,521	24.1	9.7	437,000	40.6
Truck	2,545	1.4	5.9	43,300	4.0
Pedestrian	5,244	3.0	9.0	58,000	5.4
Total	176,902	100.0	16.4	1,076,000	100.0

[a] Data supplied by the Philadelphia Urban Traffic and Transportation Board.

TABLE 18. Trends in Railroad Commutation, 1922–63[a]

Year	Passengers (In thousands)	Passenger-Miles (In millions)	Revenue Per Passenger-Mile (In cents)
1922	429,466	6,132	1.10
1923	446,538	6,401	1.09
1924	438,773	6,407	1.10
1925	446,766	6,592	1.11
1926	445,936	6,605	1.13
1927	445,171	6,650	1.11
1928	442,484	6,626	1.11
1929	457,617	6,898	1.11
1930	438,688	6,669	1.09
1931	386,349	6,018	1.06
1932	315,462	4,986	1.07
1933	271,984	4,308	1.08
1934	262,825	4,163	1.09
1935	259,099	4,112	1.09
1936	259,199	4,191	1.06
1937	245,824	4,116	1.01
1938	227,412	3,933	1.01
1939	231,126	4,012	1.02
1940	229,266	3,997	1.01
1941	232,456	4,088	1.01
1942	286,225	4,917	1.07
1943	312,246	5,261	1.07
1944	317,918	5,344	1.07
1945	322,734	5,418	1.08
1946	340,670	5,857	1.08
1947	344,604	6,008	1.12
1948	332,196	5,855	1.30
1949	308,512	5,478	1.43
1950	227,102	4,985	1.58
1951	269,464	4,866	1.71
1952	260,463	4,755	1.87
1953	255,829	4,757	1.95
1954	249,069	4,739	2.03
1955	247,759	4,776	2.12
1956	247,061	4,841	2.21
1957	249,142	4,901	2.37
1958	239,068	4,776	2.59
1959	222,486	4,549	2.75
1960	203,007	4,197	2.92
1961	198,370	4,132	3.07
1962	194,507	4,046	3.13
1963	195,081	4,101	3.17

Source: Data supplied by the Association of American Railroads. Data for 1954–63 from Association of American Railroads, *Statistics of Railroads of Class I in the United States* (Statistical Summary 48), July 1964, p. 6.

[a] Class I Railroads Revenue Passengers.

TABLE 19. Comparison of Automobile and Bus Costs
(Cents per passenger for a round trip)

| | Round Trip Mileage | | | | | | |
	1	2	3	6	10	12	20
Bus Cost:							
15¢ one-way fare	30	30	30	30	30	30	30
20¢ one-way fare	40	40	40	40	40	40	40
25¢ one-way fare	50	50	50	50	50	50	50
Hypothetical zoned fare	20	20	20	20	30	50	80
Auto Out-of-Pocket Cost (at 3.7¢ per mile):							
1 person	3.7	7.4	11.1	20.4	37.0	44.4	74.0
2 persons	1.9	3.7	5.6	10.2	18.5	22.2	37.0
3 persons	1.2	2.5	3.7	6.8	12.3	14.8	24.7
Auto Out-of-Pocket Cost plus 50¢ parking:							
1 person	53.7	57.4	61.1	70.4	87.0	94.4	$1.24
2 persons	26.9	28.7	30.6	35.2	43.5	47.2	.62
3 persons	17.9	19.1	20.4	23.5	29.0	31.5	.41
Auto Out-of-Pocket Cost plus $1.00 parking:							
1 person	$1.04	$1.07	$1.11	$1.20	$1.37	$1.44	$1.74
2 persons	.52	.54	.56	.60	.69	.72	.87
3 persons	.35	.36	.37	.40	.46	.48	.58
Total Auto Cost[a] (at 11.8¢ per mile):							
1 person	11.8	23.6	35.4	70.8	$1.18	$1.42	$2.36
2 persons	5.9	11.8	17.7	35.4	.59	.71	$1.18
3 persons	3.9	7.9	11.8	23.6	.39	.47	.79
Total Auto Cost plus 50¢ parking:							
1 person	61.8	73.6	85.4	$1.21	$1.68	$1.92	$2.86
2 persons	30.9	36.8	42.7	.61	.84	.96	$1.43
3 persons	20.6	24.5	28.5	.40	.56	.64	.95
Total Auto Cost plus $1.00 parking:							
1 person	$1.12	$1.24	$1.35	$1.71	$2.18	$2.42	$3.36
2 persons	.56	.62	.68	.86	$1.09	$1.21	$1.68
3 persons	.37	.41	.45	.57	.73	.81	$1.12

[a] Assuming all costs as enumerated in Table 22, page 125, for a car traveling 10,000 miles per year.

TABLE 20. Distance and Method of Trip to Work, Incorporated Places in Six States[a]
(In percentages)

| Distance to Work (Miles) | Workers Surveyed | Method of Transportation[b] | | |
		Auto	Public Trans-portation	Walk
0.1– 0.9	30.5	43.0	3.6	50.3
1.0– 1.9	18.7	65.8	20.1	11.8
2.0– 2.9	11.6	60.8	34.4	1.7
3.0– 4.9	14.0	58.6	35.9	0.4
5.0– 9.9	10.9	74.0	22.4	0.5
10.0–19.9	6.1	86.0	10.5	–
20 and over	3.2	84.2	6.5	–
Not reported	5.0	37.0	12.0	16.9
Total	100.0	57.6	17.5	18.7

[a] Thurley A. Bostick, Roy T. Messer, Clarence A. Steele, "Motor-Vehicle-Use Studies in Six States," *Public Roads* (December 1954), p. 111.

[b] Does not add to 100. Balance includes trips not reported and combination trips by automobile and public transportation.

TABLE 21. Highway Construction Expenditures as
Related to Traffic and Gross National
Product, 1929–54

Year	Vehicle Miles (In billions)	Highway Construction Expenditures[a] (In millions)	Highway Construction Expenditures per Vehicle-mile[a] (Cents per mile)	Highway Construction as Percentage of Gross National Product
1929	197.7	$1,978	1.00	1.2
1930	206.3	2,548	1.24	1.7
1931	216.2	2,542	1.18	1.9
1932	200.5	2,259	1.13	1.7
1933	200.6	1,567	.78	1.5
1934	215.6	1,636	.76	1.8
1935	228.6	1,448	.63	1.3
1936	252.1	2,151	.85	1.9
1937	270.1	2,064	.76	1.5
1938	271.2	2,492	.92	2.0
1939	285.4	2,478	.87	1.7
1940	302.1	2,409	.80	1.4
1941	333.4	1,739	.52	0.9
1942	267.1	931	.35	0.5
1943	206.7	516	.25	0.2
1944	211.6	461	.22	0.2
1945	248.9	526	.21	0.2
1946	340.7	1,074	.32	0.4
1947	370.6	1,532	.41	0.6
1948	397.6	1,672	.42	0.7
1949	424.1	2,128	.50	0.8
1950	458.4	2,367	.52	0.8
1951	490.4	2,349	.48	0.8
1952	512.2	2,489	.48	0.8
1953	549.7	2,856	.52	0.9
1954	557.0	3,371	.61	1.1
1956	627.8	4,301	.69	1.2
1958	664.7	5,182	.78	1.4
1960	718.9	5,004	.70	1.3
1962	767.8	5,779	.75	1.3
1963	798.0	6,113	.77	1.4

Source: U. S. Department of Commerce, Construction and Building Materials (May 1954), p. 43; the same (March 1955), p. 13; 1956–1963 figures from Automobile Facts and Figures, 1964, pp. 46 and Tables H.F.-2, Bureau of Public Roads; and GNP figures from Economic Report of the President, 1965.

[a] In 1947–49 prices.

TABLE 22. Central City Population and Area as Percentage of Total Urbanized Area

Metropolitan Area	Area in Square Miles		Percentage of Total Area Inside Central City	Percentage of Population Inside Central City
	Urbanized Area	Central City		
New York City–Northeastern New Jersey	1,892	375	20	62
Los Angeles–Long Beach, Calif.	1,370	501	37	44
Chicago–Northwestern Indiana	960	301	31	65
Detroit, Michigan	732	140	19	47
Philadelphia, Pa.–New Jersey	597	127	21	55
Baltimore, Md.	220	79	36	66
Cleveland, Ohio	587	81	14	49
St. Louis, Mo.–Ill.	323	61	19	45
Washington, D.C.–Md.–Va.	341	61	18	42
Boston, Mass.	516	48	9	29
San Francisco–Oakland, Calif.	572	101	18	46
Pittsburgh, Pa.	525	54	10	34
Milwaukee, Wis.	392	91	23	65
Houston, Texas	431	328	76	82
Buffalo, N.Y.	160	39	24	51
New Orleans, La.	267	199	75	74
Minneapolis–St. Paul, Minn.	657	109	17	58
Cincinnati, Ohio	242	77	32	51
Seattle, Wash.	238	89	37	65
Kansas City, Mo.,–Kansas	282	130	46	52
Dallas, Texas	647	280	43	73
San Diego, Calif.	276	192	70	69
Phoenix, Ariz.	248	187	75	80
Indianapolis, Ind.	145	71	49	75
Denver, Colo.	167	71	43	62
San Antonio, Texas	192	161	84	92

Source: *United States Census of Population 1960, United States Summary, Number of Inhabitants*, pp. I-40–I-49.

TABLE 23. Metropolitan Areas Crossing State Lines, 1960

Area and States Involved	Area Population	Population Opposite Side of State Line from Central City	
		Number	Per Cent
Allentown–Bethlehem–Easton (Pa. and N.J.)	492,168	63,220	12.8
Augusta (Ga. and S.C.)	216,639	81,038	37.4
Binghamton (N.Y. and Pa.)	283,600	33,137	11.7
Chattanooga (Tenn. and Ga.)	283,169	45,264	16.0
Cincinnati (Ohio, Ky. and Ind.)	1,268,479	258,117	20.3
Columbus (Ga. and Ala.)	217,985	46,351	21.3
Davenport–Rock Island–Moline (Iowa and Ill.)	319,375	119,067[a]	37.3
Duluth–Superior (Minn. and Wis.)	276,596	45,008[a]	16.3
Evansville (Ind. and Ky.)	222,890	33,519	15.0
Fall River (Mass. and R.I.)	138,156	9,461	6.8
Fargo–Moorhead (N. Dak. and Minn.)	106,027	39,080	36.9
Fort Smith (Ark. and Okla.)	135,110	47,107	34.9
Huntington–Ashland (W. Va., Ky. and Ohio)	254,780	107,601[a]	42.2
Kansas City (Mo. and Kans.)	1,092,545	329,287	30.1
Lawrence–Haverhill (Mass. and N.H.)	199,136	13,544	6.8
Louisville (Ky. and Ind.)	725,139	114,192	15.7
Memphis (Tenn. and Ark.)	674,583	47,564	7.1
Omaha (Nebr. and Iowa)	457,873	83,102	18.1
Philadelphia (Pa. and N.J.)	4,342,897	751,374	17.3
Portland (Oreg. and Wash.)	821,897	93,809	11.4
Providence–Pawtucket–Warwick (R.I. and Mass.)	821,101	89,743	10.9
St. Louis (Mo. and Ill.)	2,104,669	487,198	23.1
Sioux City (Iowa and Nebr.)	120,017	12,168	10.1
Springfield–Chicopee–Holyoke (Mass. and Conn.)	493,999	3,702	0.7
Steubenville–Weirton (Ohio and W. Va.)	167,756	68,555[a]	40.9
Texarkana (Texas and Ark.)	91,657	31,686	34.6
Toledo (Ohio and Mich.)	630,647	101,120	16.0
Washington, D.C. (Md. and Va.)	2,001,897	1,237,941	61.8
Wheeling (W. Va. and Ohio)	190,342	83,864	44.0
Wilmington (Del., N.J. and Md.)	414,565	107,119	25.8

Area and States Involved	Area Population	Population Opposite Side of State Line from Central City	
		Number	Per Cent
[Standard Consolidated Areas] New York–Northwestern N.J.	14,759,428[b]	4,064,796	27.5
Chicago, Ill.–Northwestern Indiana	6,794,461[c]	573,548	8.4
Total	41,119,583		

Source: Bureau of the Budget, *Standard Metropolitan Statistical Area,* 1964, pp. 4–43.

[a] Where "central cities" exist on both sides of a state line, the cities with the smaller population totals have been considered for the purpose of this tabulation to be "on the opposite side of the state line."

[b] Includes Standard Metropolitan Statistical areas of New York, Newark, Jersey City, and Paterson–Clifton–Passaic, and Middlesex County and Somerset County in New Jersey.

[c] Includes Standard Metropolitan Statistical areas of Chicago and Gary–Hammond–East Chicago in Indiana.

Index